Essays in
Biochemistry

volume 33 1998

Essays in Biochemistry

Molecular Biology of the Brain

Edited by S.J. Higgins

84, 651

PORTLAND PRESS

Essays in Biochemistry is published by Portland Press Ltd
on behalf of The Biochemical Society

Portland Press Ltd
59 Portland Place
London W1N 3AJ
U.K.
fax: 0171 323 1136; e-mail: edit@portlandpress.co.uk

Published in North America by Princeton University Press,
41 William Street, Princeton, NJ 08540, U.S.A.

British Library Cataloguing-in-Publication Data
A catalogue record for this book is available from the British Library

ISBN   1 85578 086 0
ISSN   0071 1365   2079

Typeset by Portland Press Ltd
Printed in Great Britain by Information Press Ltd, Eynsham, U.K.

Contents

Preface..xi

Authors..xiii

Abbreviations...xix

1 Molecular cues that guide the development of neural connectivity
Guy Tear

Introduction I
Netrins and their receptors..3
Semaphorins/collapsins and their receptors....................................5
Eph receptor tyrosine kinases and their ligands, the ephrins.......6
Extracellular matrix (ECM) and cell adhesion molecules 8
Guidance molecules at intermediate targets...................................9
Signal transduction of axon guidance signals 9
Future perspectives...10
Summary..11
References..11

2 Understanding neurotransmitter receptors: molecular biology-based strategies
Mark Wheatley

Introduction 15
How many receptors?..16
Approaches to determining the structures and physiological
functions of native neurotransmitter receptors...........................19
Receptor architecture and function...23
Future perspectives...25
Summary..26
References..26

3   Molecular analysis of neurotransmitter release
Giampietro Schiavo and Gudrun Stenbeck

Introduction..29
SNARE proteins..31
Synaptotagmins as Ca^{2+} sensors at the synapse.............................35
Rab3 and neurotransmitter release 36
Phosphoinositide biosynthesis and turnover at the nerve terminal........37
Cytoskeleton and exocytosis 38
Proteins involved in synaptic vesicle endocytosis.........................39
Perspectives..39
Summary...40
References..41

4   Mitochondria in the life and death of neurons
Samantha L. Budd and David G. Nicholls

Introduction..43
Bioenergetic functions of brain mitochondria 44
Mitochondria and excitotoxicity..45
Mitochondrial hypotheses for neurodegenerative disorders...................46
Mitochondria and programmed cell death 49
Perspectives..50
Summary...51
References..51

5   Neuro-regeneration: plasticity for repair and adaptation
Pico Caroni

Introduction..53
Reactions of adult neurons to axotomy 55
Role of extrinsic factors in axonal regeneration...........................57
Role of intrinsic neuronal components in axonal regeneration 59
Factors that control nerve sprouting and synaptogenesis in the adult....60
Perspectives..62
Summary...63
References..63

6   A molecular basis for opiate action
Dominique Massotte and Brigitte L. Kieffer

Introduction..65
The opioid system: discovery of a complex neurotransmitter system....67

Opioid receptors: first steps towards molecular mechanisms
of opioid action..70
Recent approaches: from gene to function...............................74
Perspectives: an arduous march to therapeutics....................75
Summary...76
References..76

7 Gases as neurotransmitters
Jane E. Haley

Introduction...79
Why is everyone so interested in these gases?.......................80
NO and CO are formed by enzymes...80
What are the targets for NO and CO?..82
LTP in the hippocampus 83
Do NO and CO contribute to nociceptive signalling within
the spinal cord?...88
Summary...89
References..90

8 Molecular biology of olfactory receptors
Yitzhak Pilpel, Alona Sosinsky and Doron Lancet

Overview..93
Chemical detection in a probabilistic receptor repertoire 93
ORs belong to the G-protein-coupled receptor (GPCR) hyperfamily....94
Odorant complementarity-determining regions.....................96
Evolution of the OR repertoire...96
The biochemical cascade in olfactory signalling......................96
Expressed OR proteins and their ligand specificity................98
Patterns of olfactory receptor expression and their transcriptional
regulation...99
Olfactory bulb glomeruli represent ORs....................................101
Summary...102
References 102

9 Pathology and drug action in schizophrenia: insights from molecular biology
Philip G. Strange

Introduction...105
Schizophrenia: the clinical picture...105
Changes in the brain in schizophrenia.......................................106

Genetic linkage analysis of schizophrenia 108
Drug action in schizophrenia 109
Inverse agonism of anti-psychotic drugs 113
Conclusions 114
Summary 115
References 115

10 Genetics of Alzheimer's disease

Michael Hutton, Jordi Pérez-Tur and John Hardy

Introduction 117
Early-onset, autosomal dominant disease: the amyloid precursor
protein (APP) and the presenilins 118
Effect of presenilin mutations on Aβ42(43) 123
Early-onset autosomal dominant AD 125
ApoE and other genetic risk factors for AD 126
Are there other Alzheimer genes, and what are they likely to be? 127
Summary 128
References 129

11 Use of brain grafts to study the pathogenesis of prion diseases

Adriano Aguzzi, Michael A. Klein, Christine Musahl, Alex J. Raeber, Thomas Blättler, Ivan Hegyi, Rico Frigg and Sebastian Brandner

Introduction 133
Biological characteristics of mouse neuroectodermal grafts 135
Blood–brain barrier and brain grafts 135
Neurografts in prion research 137
Spread of prions in the CNS 138
Cells in the CNS that are affected by spongiform encephalopathies 141
Summary 145
References 145

12 Pathological mechanisms in Huntington's disease and other polyglutamine expansion diseases

Astrid Lunkes, Yvon Trottier and Jean-Louis Mandel

Introduction 149
HD: clinical features 150
HD: neuropathology 154
Expansion mutation and genotype–phenotype correlations 154
Polyglutamine expansions in other neurodegenerative disorders 155
Nuclear inclusions and mechanisms of neurodegeneration 157

Summary......161
References 161

13 The matter of mind: molecular control of memory
Emily P. Huang and Charles F. Stevens

Introduction......165
Synaptic basis of memory......166
LTP 167
Mutant mice, memory and LTP 170
Drosophila memory and cAMP 172
Long-term memory in *Drosophila* 173
CREB and mammalian memory 174
Perspectives......176
Summary......176
References 177

14 Future developments
Susan Greenfield

Introduction......179
Why are there so many different neurotransmitters? 180
Why should neurotransmitters be released from outside of
the classical synapse? 182
How can familiar transmitters have unpredictable actions?......185
How do transmitter actions relate to function? 186
How do transmitters relate to dysfunction? 187
Conclusions 189
Summary......189
References 190

Subject index 193

Preface

The workings of the brain have long been a fascination for scientists. Yet, faced with the obvious anatomical and biochemical complexity of the brain, understanding its functions more than superficially seemed an impossible goal. Equipped with no more sophisticated analytical techniques than, for instance, 'grind and find', early biochemists could do little more than identify neurotransmitters, work out the basic processes of neurotransmission and provide simplistic explanations for the neuropharmacological actions of simple drugs. Clearly this state of affairs was woefully inadequate to provide biochemical explanations for pain, pleasure or addiction, or to tackle the major neurological disorders, let alone scratch the surface of phenomena such as mood, anxiety or memory.

The advent of the Molecular Biology Era has revolutionized our research technology, and with it our thinking. Our ability to isolate and manipulate specific genes, identify their protein products and determine their functions has given us unprecedented power to analyse complex biological systems, and nowhere are the fruits of this more obvious than in brain research. The brain is now no longer *Terra Incognita* but *The New Frontier*.

The authors of the essays in this volume, acknowledged experts in their specialties, have illustrated the power of molecular biology to dissect the molecular functioning of the brain. The early part of the volume has related essays on neurotransmitters and their receptors by Mark Wheatley, Giampietro Schiavo and Gudrun Stenbeck, and Jane Haley. Aspects of neuronal development and neurodegeneration are discussed by Guy Tear, Pico Caroni, and Samantha Budd and David Nicholls. The molecular biology of opiate action (Dominique Massotte and Brigitte Kieffer) is crucial to understanding these drugs and their addictive properties, but also to explaining natural analgesia. The concept of neuronal networks is nicely illustrated for the olfactory system by Doron Lancet's group. Neurodegenerative and affective disorders are major healthcare problems that can be expected to yield to analysis by molecular genetical approaches. In these areas we have essays by Philip Strange (schizophrenia), John Hardy's group (Alzheimer's disease), Adriano Aguzzi et al. (spongiform encephalopathies; prion diseases) and Jean-Louis Mandel and colleagues (trinucleotide expansion disorders). Emily Huang and Chuck Stevens then provide an exciting account of how molecular biology is beginning to explain a phenomenon as complex as memory. The volume ends with a thought-provoking essay on Future Developments by Susan Greenfield.

The brief given to authors was to convey the excitement of their field, point to future developments, and encourage their audience — undergraduates in their senior years and starting postgraduates — to want to become part of the research effort that will open up the *New Frontier*. I think that they will succeed — so after reading these essays all that will remain to be done is to *Go West Young (Wo)Man!*

Steve Higgins
University of Leeds, 1998

Authors

Guy Tear is an MRC Senior Research Fellow in the Department of Biochemistry at Imperial College, London. His current research is an investigation of the molecules required for axon guidance during the development of the Drosophila central nervous system. Before moving to Imperial College, he received his PhD from Cambridge University and undertook a period of postdoctoral research at the University of California, Berkeley.

Mark Wheatley graduated from the University of London with a BSc in Biochemistry in 1980. His studies on the D_2 dopamine receptor in the laboratory of Philip Strange resulted in the degree of PhD in Biochemistry being awarded by the University of Nottingham in 1983. Between 1983 and 1987 he was a Research Fellow at the National Institute for Medical Research in London, working with Drs Ed Hulme and Nigel Birdsall on the structure and function of the muscarinic acetylcholine receptor. In 1988 he was appointed to a lectureship in Biochemistry at the University of Birmingham. He is currently Senior Lecturer in Biochemistry and his research interests address the structure and function of the G-protein-coupled receptors for peptide hormones, particularly vasopressin and oxytocin.

Giampietro Schiavo graduated in 1992 from the laboratory of C. Montecucco, University of Padua, Italy, with a thesis on the effect of tetanus and botulinum neurotoxins on neurotransmitter release. He is now pursuing his interest in the mechanism of neurosecretion as group leader at the Imperial Cancer Research Fund in London. Gudrun Stenbeck graduated in 1993 from the University of Heidelberg, Germany (F. Wieland's laboratory), with a thesis on the characterization of the coatomer subunits of COPI-coated vesicles involved in Golgi transport. She is now a lecturer at University College London. Her interest is in the molecular mechanism of bone remodelling. Both authors had postdoctoral training, on the molecular basis of neurotransmitter release, in the laboratory of J.E. Rothman, at the Memorial Sloan-Kettering Cancer Center in New York.

Samantha Budd is currently undertaking postdoctoral research at the CNS Research Institute, Brigham & Women's Hospital, Harvard Medical School, Boston. She completed her PhD with David Nicholls in Dundee in 1997. David Nicholls is Professor of Neurochemistry in the Neurosciences Institute of the University of Dundee. He has a long-standing interest in mitochondrial function, including brown fat uncoupling, Ca^{2+} transport and their *in situ* function in isolated nerve terminals and cultured neurons. He is the author of *Proteins, Transmitters and Synapses* (published by Blackwell Science) and co-author of *Bioenergetics 2* (published by Academic Press). His current interest is in mitochondrial (dys)function during excitotoxicity.

Pico Caroni is a staff member of the Friedrich Miescher Institute (FMI, Basel, Switzerland), an institute supported by Novartis that is dedicated to basic research in biology. He obtained a PhD in biochemistry at the ETH in Zürich under Ernesto Carafoli, and was a postdoctoral fellow in Regis Kelly's group at UCSF. In 1985 he joined Martin Schwab's group at the Brain Research Institute of the University of Zürich, where he established the existence of oligodendrocyte components that inhibit axonal growth. In 1989 he moved to the FMI, where his research interests focus on structural plasticity in the adult nervous system.

Dominique Massotte performed her doctoral work on structure–function studies of membrane proteins in Liège, Belgium, and at the European Molecular Biology Laboratory in Heidelberg, Germany, and graduated from Liège. She received a postdoctoral fellowship at the Max-Planck Institute in Cologne (Germany) where she trained in molecular biology and developed the baculovirus expression system. She now holds a CNRS research position at the Ecole Supérieure de Biotechnologie de Strasbourg (France) where she initiated a structure–function study of human opioid receptors. **Brigitte Kieffer** graduated in Chemistry and Biochemistry from the University of Strasbourg, France, and was recipient of a postdoctoral fellowship at the Friedrich Miescher Institute in Basel, Switzerland, where she trained in molecular biology. She started her own research as an associate professor at the Ecole Supérieure de Biotechnologie de Strasbourg where she initiated the molecular biology of opioid receptors. She now is a professor at the Faculty of Pharmacy (France) where she runs a genetic engineering teaching programme, and leads a research group studying opioid receptor function at the Ecole Supérieure de Biotechnologie de Strasbourg.

Jane Haley studied for a BSc in Pharmacology at University College London (UCL) and worked for Wellcome upon graduation. She returned to UCL in 1987 to undertake a PhD, during which she became interested in a possible role for nitric oxide in longer-duration pain states and spinal hyperalgesia. She continued this work at the University of Minnesota, U.S.A., and extended her research to include the involvement of nitric oxide in the induction of long-term potentiation (LTP). She moved to Stanford University in 1993 where she also investigated carbon monoxide involvement in LTP. She is currently a postdoctoral researcher at UCL examining G-protein modulation of ion channels.

Yitzhak Pilpel obtained his degree in Biology from the Tel Aviv University, Israel, in 1993, and is now studying towards a PhD in the Department of Molecular Genetics at the Weizmann Institute of Science, Rehovot, Israel. He is currently researching bioinformatics and molecular modelling of the olfactory receptor proteins. **Alona Sosinsky** graduated in Biophysics from the St Petersburg Technical University, Russia, in 1993, and is now studying towards a PhD in the Department of Molecular Genetics at the Weizmann Institute of Science. His current research interests are centred on gene structure and expression of olfactory genes. **Doron Lancet** obtained a BSc in Chemistry and Physics at the Hebrew University, Jerusalem, Israel, in 1970. He studied towards his PhD at the Weizmann Institute of Science and undertook postdoctoral research at Harvard University with Jack Strominger. There followed a period as a postdoctoral research associate at Yale University School of Medicine, with Gordon Shepherd. He is currently a professor in the Department of Molecular Genetics at the Weizmann Institute of Science with interests in molecular biology, genetics, human genomics of the sense of smell, molecular recognition and the origin of life. He is also Head of the Genome Center at the Weizmann Institute.

Philip Strange graduated from Cambridge University with a BA in Chemistry in 1970 and a PhD in Organic Chemistry in 1973. Following postdoctoral research in Berkeley, California (D.E. Koshland's laboratory), and Mill Hill, London (A.S.V. Burgen's laboratory), he has held positions at the universities of Nottingham (Lecturer in Biochemistry) and Kent (Reader in Biochemistry and then Professor of Neuroscience) before taking up his current position as Professor of Neuroscience at the University of Reading in 1998.

John Hardy was born in Nelson, Lancashire (1954) and spent his childhood in Lancashire and Cheshire in northern England. He completed his degree in Biochemistry at Leeds University in 1976 and his PhD in Neurochemistry at Imperial College in 1979. After postdoctoral positions in Newcastle upon Tyne (1979–1983) and Umea, Sweden (1983–1985), he returned to Imperial College as a Lecturer (1985), then a Senior Lecturer (1989), before taking a Chair at the University of South Florida in 1992. Since 1995 he has been at the Mayo Clinic, Jacksonville, Florida. **Michael Hutton** completed his undergraduate training at the University of Manchester (1985–1988) before moving on to do a PhD in Molecular Neuroscience at the University of Cambridge (1989–1992) working in the laboratory of Eric Barnard on ligand-gated ion channels. After completing postdoctoral training in Cambridge (1992–1994), he joined John Hardy's group at the University of South Florida, and moved with him to the Mayo Clinic in 1996. Hutton now runs his own laboratory, having been appointed as a Senior Associate Consultant and Assistant Professor in 1997. His current research focuses on the role of the Tau protein in neurodegeneration. **Jordi Perez-Tur** obtained his degree in Biology in 1987 from the Universitat de Valencia-Estudi General, Spain, and his PhD (Molecular Biology) from Universidad Autonoma de Madrid, Spain, in the Centro de Biologia Molecular in 1993. He moved to his first postdoctoral stage in 1993, when he joined the group of Marie-Christine Chartier at INSERM, Lille, France. His second postdoctoral position (1995–1996) was under the direction of John Hardy at the University of South Florida, and in 1996 he moved with Hardy to the Mayo Clinic. In 1998 he was appointed Assistant Professor and has been recently appointed Associate Consultant.

Adriano Aguzzi is Professor for Neuropathology at the Institute of Neuropathology, University Hospital Zurich, Switzerland. He is leading a research group of around 20 co-workers, who are involved in work on the pathogenesis of prion diseases and neurocarcinogenesis. Among the people in the 'prion group' are **Michael A. Klein, Ivan Hegyi** and **Rico Frigg**, all of whom have a Masters degree. Their projects focus on the peripheral pathogenesis and lymphotropism of prion diseases. **Thomas Blättler** finished his thesis on the transfer of scrapie infectivity from spleen to brain in 1996. **Alex Raeber** started his scientific career in the laboratory of Stanley Prusiner, before he joined Charles Weissmann in 1993. He is now senior postdoc in the laboratory and is supervising a group of several PhD students. **Sebastian Brandner** joined Adriano Aguzzi in 1993. He established the neurografting technique and used it for the study of prion neurotoxicity. **Christine Musahl**, a postdoctoral fellow since 1997 in the laboratory, established a transgenic mouse model to study the role of PrP-specific antibodies in the pathogenesis of prion diseases.

Astrid Lunkes obtained her PhD in 1995 at the University of Düsseldorf, Germany, working on the positional cloning of spinocerebellar ataxia 2. In 1996 she joined the laboratory of Jean-Louis Mandel to investigate functional aspects involved in the pathogenesis of Huntington's disease using a cellular model approach. **Yvon Trottier** graduated from Laval University (Québec), investigating the role of cytochromes P-450IA in chemical carcinogenesis in human cells. Since 1992, she has been studying the mechanism whereby polyglutamine expansion causes neurodegenerative disorders, like Huntington's disease and spinocerebellar ataxia. **Jean-Louis Mandel** is Professor of Genetics at the Faculty of Medicine of Strasbourg and Head of Human Molecular Genetics at the Institut de Génétique et Biologie Moléculaire et Cellulaire, Strasbourg. In 1982, he initiated a project on mapping and identification of disease genes, leading notably to the discovery, in 1991, of unstable mutations in the fragile X mental retardation syndrome, the isolation of the gene responsible for adrenoleukodystrophy, and, most recently, to the cloning of the myotubular myopathy gene. Since 1991, he has been involved in the analysis of diseases caused by trinucleotide repeat expansions (including Huntington's disease, spinocerebellar ataxia and Friedreich ataxia).

Emily P. Huang is a research associate with the Howard Hughes Medical Institute and Molecular Neurobiology Laboratory at the Salk Institute in La Jolla, California, U.S.A. **Charles F. Stevens** is a Howard Hughes Medical

Institute Investigator and Professor of Molecular Neurobiology at the Salk Institute in La Jolla, California, U.S.A.

Susan Greenfield read for a first degree at St Hilda's College, Oxford, and worked for a DPhil in the University Department of Pharmacology. She held postdoctoral fellowships in Oxford, Paris and New York, until being appointed, in 1985, as University Lecturer in Synaptic Pharmacology and Fellow and Tutor in Medicine, Lincoln College. Since then she has also held a Visiting Research Fellowship at the Institute of Neuroscience, La Jolla, and was the 1996 Visiting Distinguished Scholar, Queens University, Belfast. The title of Professor of Pharmacology was conferred in 1996. In 1997 she was awarded an Honorary DSc by Oxford Brookes University, and is to receive Honorary DSc degrees, in 1998, from the University of St Andrew's and Exeter University. She became Director of The Royal Institution of Great Britain in 1998. Apart from her primary research, where she heads a multidisciplinary group studying how diverse neurons prone to degeneration share a common yet non-classical feature. Greenfield has developed an interest in the physical basis of the mind and has edited or authored a number of publications on the subject.

Greenfield also makes contributions to the public understanding of science. In 1994 she was the first woman to be invited to give the Royal Institution Christmas lectures and has subsequently made a wide range of broadcasts on TV and radio. She is currently preparing a major six-part series on the brain and mind, to be broadcast in the year 2000. In 1995 she was elected to the Gresham Chair of Physics. She was general editor in 1996 for *The Human Mind Explained* (Cassell) and has recently authored *The Human Brain: A Guided Tour* (Weidenfeld & Nicholson) which reached the bestseller list. In addition, she writes a fortnightly column for *The Independent on Sunday* on aspects of science, as well as contributions to *The Times, The Times Higher Education Supplement, The Sunday Times, The Independent* and *The Daily Telegraph*. She was ranked by Harpers and Queen as number 14 in the '50 Most Inspirational Women in the World'.

Abbreviations

$\Delta\psi$	membrane potential
ACh	acetylcholine
AChR	acetylcholine receptor
AD	Alzheimer's disease
AIF	apoptosis-inducing factor
AMPA	α-amino-3-hydroxy-5-methylisoxazolepropionate
AP5	amino-5-phosphonopentanoate
ApoE	apolipoprotein E
APP	amyloid precursor protein
BBB	blood–brain barrier
BSE	bovine spongiform encephalopathy
CAM	cell adhesion molecule
CaMKII	Ca^{2+}/calmodulin-dependent kinase
CAP-23	cortical cytoskeleton-associated protein of 23 kDa
CJD	Creutzfeldt–Jakob disease
CNS	central nervous system
comm	commissureless
CREB	cAMP response element binding protein
DCC	deleted in colorectal cancer
DRG	dorsal root ganglion
DRPLA	dentatorubral–pallidoluysian atrophy
D_{2S}, D_{2L}	short and long forms respectively of the D_2-dopamine receptor
ECM	extracellular matrix
EDRF	endothelium-derived relaxing factor
eNOS	endothelial NOS
GABA	γ-aminobutyric acid
GAP	growth-associated protein
GC	guanylate cyclase
GFAP	glial fibrillary acidic protein
GluR	glutamate receptor
GPCR	G-protein-coupled receptor
GPI	glycosylphosphatidylinositol
HD	Huntington's disease
HO	haem oxygenase
5-HT	5-hydroxytryptamine
i3	third intracellular loop
IgCAM	immunoglobulin CAM
iNOS	inducible NOS
KO mice	knock-out mice
LANP	leucine-rich acidic nuclear protein

LGIC	ligand-gated ion channel
LTP	long-term potentiation
mAChR	muscarinic AChR
MJD	Machado–Joseph disease
MPP$^+$	1-methyl-4-phenylpyridinium
MPTP	1-methyl-4-phenyl-1,2,3,6-tetrahydropyridine
MRI	magnetic resonance imaging
mtDNA	mitochondrial DNA
NK$_2$R	tachykinin NK$_2$ receptor
N-CAM	neural CAM
NMDA	N-methyl-D-aspartate
NOS	nitric oxide synthase
nNOS	neuronal NOS
NSF	N-ethylmaleimide-sensitive factor
nvCJD	new variant CJD
OR	olfactory receptor
Δp	proton motive force
PD	Parkinson's disease
PET	positron emission tomography
PKA	cAMP-dependent protein kinase
PNS	peripheral nervous system
POMC	prepro-opiomelanocortin
PRA	prenylated Rab acceptor
PrP^C	normal prion protein
PrP^{Sc}	pathologically changed isoform of PrP^C
PS-1(-2)	presenilin-1(-2)
PSD	postsynaptic density protein
R, R*	inactive and active receptor forms respectively
RAGS	repulsive axon guidance signal
robo	roundabout
SBMA	spinal and bulbar muscular atrophy
SCA	spinocerebellar ataxia
SH2	Src homology 2
SNAP	soluble NSF accessory protein
SNAP-25	synaptosomal-associated protein of 25 kDa
SNARE	SNAP receptor
SSV	small synaptic vesicle
t-SNARE	target membrane SNARE
TM	transmembrane
VAMP	vesicle-associated membrane protein
v-SNARE	vesicular SNARE

1

Molecular cues that guide the development of neural connectivity

Guy Tear

Department of Biochemistry, Imperial College of Science, Technology and Medicine, Exhibition Road, London SW7 2AZ, U.K.

Introduction

The nervous system is composed of a very large number of interconnecting neurons. This intricate pattern must be assembled precisely to ensure that the brain can perform its remarkable functions of cognition and behaviour. Observations over a period of a century have led to the identification of a number of the principles that underlie these feats of navigation, while recent advances have unearthed the molecules responsible [1]. We are now entering an era where we can begin to describe in molecular terms how the brain is constructed.

The pattern of neural connectivity takes shape during embryonic development, once the organism has established its population of neuronal cells. These cells extend processes that must navigate from their site of birth through the rich extracellular environment to find their appropriate targets, which can be a large distance away. At the tip of the advancing neuron is the growth cone, which extends finger-like extensions (filopodia) that sample the immediate environment and react to the cues they encounter. These processes withdraw from unfavourable cues, while maintaining and growing towards favourable contacts. In this manner the first neurons to extend pioneer a number of pathways that provide a scaffold which later axons can follow and build upon. The follower neurons themselves have to be able to make a choice of which of the many pre-existing pathways to join, and must also decide when to branch off again to reach their specific target. Over the last decade or so the combination

of improved biochemical assays and molecular genetic approaches in vertebrates and invertebrates has begun to identify the molecular nature of the cues
that are responsible for directing growth cones. It has become clear that the
growth cone is guided by molecules that function as attractants to direct neurons towards their target, or as repellents that direct neurons away from inappropriate territories. The molecules may be diffusible and able to act over long
distances, or may be associated with cell surfaces or the extracellular matrix
and act over short distances (Figure 1). Furthermore, different forms of the
same type of molecule can act either as attractants or as repellents. Many of
these extracellular molecular cues act as ligands which bind receptors on specific neurons to activate a signal transduction pathway that in turn directs
extension of the growth cone in the appropriate direction. In this essay I will
consider some of the molecules that guide particular neurons, the nature of
their receptors and the mechanisms that signal the appropriate growth-cone
response. The molecular strategies identified so far appear to make full use of
the coding capacity of the genome to direct the error-free growth of billions of
neurons along their defined pathways.

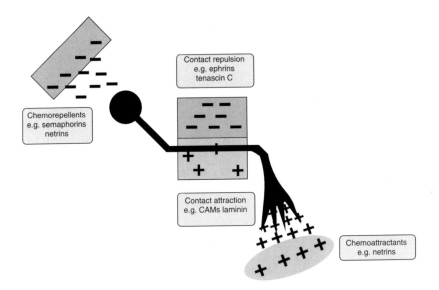

Figure 1. Types of axon guidance cue
Four types of guidance cue have been identified that direct the projections taken by axonal
growth cones. The cues can either be secreted (chemorepellents or chemoattractants) and influence the axon at a distance from their site of synthesis, or be provided by the surface that the
axon contacts (contact-mediated repulsion or attraction). Chemorepellents repel growth cones
from extending along an inappropriate projection, while chemoattractants orient the growth
cone towards an intermediate or final target. Contact-mediated cues are either attractive and
provide a permissive surface for extension, or repellent and inhibit extension. Examples of molecules responsible for these cues are provided (note that individual molecules can be bifunctional).
The integration of the different signals provided by these cues ensures accurate axon guidance
during neural development.

Netrins and their receptors

One common mechanism by which pioneer axons simplify the task of navigating large distances is to make use of intermediate targets or choice points along the pathway towards their target. An intermediate target found in all bilaterally symmetrical organisms is the midline, which separates the two halves of the nervous system. Cells at the midline provide both attractive and repulsive guidance cues to ensure that appropriate axons, such as the commissural axons that link the two sides, are directed to the midline, while keeping other axons away. Co-culture experiments demonstrate that floor-plate explants from the midline of the vertebrate neural tube provide chemotropic guidance cues that attract commissural axons. Purification of this activity from embryonic chick brain led to the identification of the netrins [2]. Mouse netrin-1 is expressed at high levels at the floor plate and less strongly in the ventral two-thirds of the spinal cord. This results in a graded expression of netrin, with a high point at the floor plate, which the commissural axons follow to reach the ventral midline. Consistent with this is the observation that, in *netrin-1* mutant mice, the majority of commissural axons fail to reach the ventral midline [3].

The netrins are secreted proteins of about 600 amino acids which include an N-terminal domain similar to domains VI and V of laminin and a basic C-terminal domain that is conserved within the netrins (Figure 2). Netrin homologues exist in the nematode (UNC-6; UNC, uncoordinated) and in *Drosophila* (netrins A and B), where they are also expressed at the ventral midline of the organism. Furthermore, loss-of-function mutations in these organisms reveal a conserved role for netrins in directing the ventral projection of axons [1].

Genetic analysis in the nematode revealed that mutations in genes encoding netrins also affect the guidance of dorsally projecting axons that head away from the midline. This led to the realization that netrin performs a dual role and is also able to repel neurons [4]. Further evidence for a dual role came from experiments *in vitro* in which tissue culture cells secreting netrin-1 were capable of repelling vertebrate motor axons that normally extend away from the floor plate *in vivo* [5].

The differential response to netrins depends on the nature of the netrin receptor expressed by axons. The receptors responsible were initially identified by genetic analysis in the nematode. The first of these, UNC-40, is homologous to the protein encoded by the vertebrate gene *deleted in colorectal cancer* (DCC), and is the netrin receptor responsible for directing axons to the midline. DCC encodes a transmembrane protein containing four immunoglobulin domains and six fibronectin type III repeats (Figure 2) that is expressed on the growth cones of commissural axons. Mice lacking a functional DCC display defects in commissural axon projections that are similar in nature to, yet more severe in outcome than, those in netrin-1-deficient mice [6]. Similarly, mutations in the *Drosophila* homologue of DCC, *frazzled*, lead to defects in axonal projections towards the midline [7].

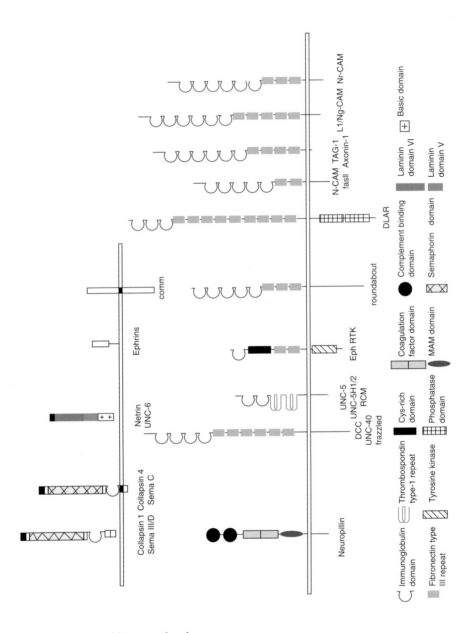

Figure 2. Axon guidance molecules

Representatives of the different families of molecules that have a role in axon guidance are shown. The semaphorins (Sema) bind to neuropilin, netrin binds DCC (deleted in colorectal cancer) and UNC-5H1/2 and the ephrins are the ligands for the Eph receptor tyrosine kinases (RTKs). Most of the axon guidance molecules have a modular construction, often containing tandem arrays of domains previously identified in other proteins, as indicated. Abbreviations: comm, commissureless; RCM, rostral cerebellar malformation; DLAR, *Drosophila* leukocyte antigen related ; N-CAM, neural cell adhesion molecule; fasII, fasciclin II; TAG, transient axonal glycoprotein; Ng-CAM, neuron–glia cell adhesion molecule; Nr-CAM, NgCAM-related cell adhesion molecule.

UNC-5 is the nematode receptor that may function to interpret the netrin signal as a chemorepellent. Three vertebrate homologues of this molecule have been identified, UNC-5H1, UNC-5H2 and rostral cerebellar malformation (RCM), each of which are able to bind netrin-1 [8]. All these molecules are transmembrane proteins and have two immunoglobulin domains and two thrombospondin domains (Figure 2). Their expression patterns are suggestive of a role in axonogenesis. It is unclear whether these molecules function as receptors themselves or in a complex with UNC-40/DCC to mediate a differential response to the netrin signal.

Semaphorins/collapsins and their receptors

The semaphorins/collapsins are a recently characterized family of molecules that mediate axon guidance through inhibition of outgrowth. The identification of molecules that are able to act as repellents has been predicted by the observation of growth-cone behaviour. For example, growth cones of sympathetic neurons collapse on contacting retinal axons, and chemorepellent activities divert olfactory bulb axons away from the septum. The collapsin 1 molecule was purified as a factor derived from chick brain that is able to cause the collapse of growth cones that extend from explants of the dorsal root ganglion (DRG) [9]. Subsequently it was observed that growth cones will turn away from beads coated with collapsin 1, demonstrating that these molecules can also direct axonal growth. Many further members of this new family, the semaphorins, were rapidly identified. The family was found to include both secreted and membrane-bound forms that share a common domain of approx. 500 amino acids, the semaphorin domain (Figure 2). The different semaphorins have varying expression patterns, suggesting a role in the formation of a number of axon pathways. In this role the semaphorins are able to act selectively; thus collapsin 1/semaphorin III will inhibit the outgrowth of nerve-growth-factor-dependent DRG sensory neurons, while having little or no effect on neurotropin-3-responsive neurons [10]. A 70-amino-acid region within the semaphorin domain appears to be sufficient to specify the selective biological activity of the different semaphorins [11]. That the semaphorins act as axon repellents *in vivo* is also suggested by *semaIII* mutant mice, which display abnormal projections, including extension of neurons beyond their targets into regions that normally express high levels of semaphorin III [12].

Recently, members of the neuropilin family have been identified as candidate receptors for the semaphorins, since they are able to bind semaphorins with high affinity [13,14]. However, further components are likely to be required to provide receptor specificity, since studies *in vitro* suggest that while both semaphorin III and semaphorin E can bind to neuropilin-1, only the former can cause the collapse of nerve-growth-factor-dependent DRG sensory neurons that express neuropilin-1. Furthermore the neuropilin cytoplasmic domain is small, about 40 amino acids in length, with no obvious motifs, suggesting that further molecules are required to co-operate in the transduction of the semaphorin signal.

Eph receptor tyrosine kinases and their ligands, the ephrins

The development of the highly ordered axonal projections linking specific areas of the retina to specific areas of the optic tectum is also mediated by growth-cone repulsion. Axons from the temporal retina project to the anterior tectum, and axons from the nasal retina project to the posterior tectum, such that an inverted topographical map of the innervating axons is formed on the tectum (Figure 3a). In an *in vitro* assay, temporal retinal axons prefer to extend

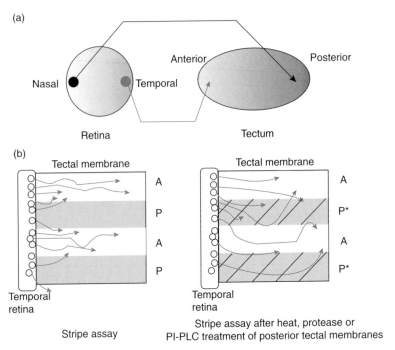

Figure 3. Retinotectal axon guidance
(a) Axons project from the retina to form a topographical map on their target tissue, the tectum.
(b) When temporal retinal axons are allowed to grow out on to strips of posterior (P) or anterior (A) tectal membrane, they extend preferentially on the anterior tectum. This preference is abolished when posterior tectal membranes are subject to treatments that result in the removal of GPI-linked proteins (P*). This reveals that the posterior tectal membranes contain GPI-linked proteins that repel temporal axons. This inhibitory activity is provided by ephrin-A2 and -A5. (a) Ephrin-A2 (grey shading) is expressed in an increasing anterior-to-posterior gradient across the tectum (ephrin-A5 is expressed in a similar gradient, but is restricted to the posterior portion of the tectum). The Eph receptor EphA3, which binds both ephrin-A2 and -A5, is expressed in a reciprocal gradient across the retina (blue shading). Such an arrangement of reciprocal gradients of ligand and receptor may provide the appropriate information to allow formation of the highly ordered projection pattern in this system, in which neurons with low levels of receptor can extend into regions expressing high levels of the repellent ligand. Abbreviation: PI-PLC, phosphoinositide-specific phospholipase C.

on the anterior tectum membrane and steer away from the posterior tectum membrane (Figure 3b). Pretreatment of the tectum membrane to remove glycosylphosphatidylinositol (GPI)-anchored proteins overcomes this preference, and temporal axons will then extend on both the anterior and the posterior tectum. Purification of a developmentally regulated GPI-anchored protein enriched in the posterior tectum identified the RAGS (repulsive axon guidance signal) molecule [15]. Characterization of RAGS revealed that it is a member of the ephrin family of ligands for the Eph family of receptor tyrosine kinases. RAGS (renamed ephrin-A5) is able to induce growth-cone collapse of temporal axons. The possibility that ephrin-A5 provides topographical information is suggested by its differential expression along the anterior–posterior axis of the tectum to form a gradient with a high point at the posterior. A second ephrin, ELF-1 (Eph ligand family; ephrin-A2), is expressed in a similarly graded fashion in the tectum of the mouse and chick (Figure 3a). Ectopic patches of ephrin-A2 in the chick tectum, produced using a retrovirus, are avoided by temporal axons, suggesting that these molecules repel growth cones *in vivo* [15]. An ephrin-A2 receptor, EphA3 (Mek4), is expressed in retinal ganglion cells, where it forms a gradient with a high point of expression on the temporal side. Thus retinal ganglion cells with high levels of receptor (temporal) connect to tectal regions with low levels of the inhibitory ligand (anterior) (Figure 3a). This leads to a simple model in which each axon seeks out a specific concentration of ligand depending on its level of receptor expression. However, it seems unlikely that all the properties of retinotectal map formation, such as axon retraction and branching and the plasticity observed in size-disparity experiments, can be explained by this simple model. Further complexity is already evident: EphA3 binds ephrin-A5 and -A2 with different affinities; the gradient of ephrin-A5 expression is steeper than that of ephrin-A2 and more confined to the posterior tectum; and ephrin-A5 can produce a concentration-dependent guidance of nasal and temporal axons, whereas ephrin-A2 only affects temporal axons [16].

Eph receptor molecules are also expressed on subsets of developing motor axons, and ephrins are found in target areas such as the limb buds, suggesting a widespread role for these molecules in axon guidance. The Eph receptors and the ephrins also play a role in regulating the formation of axon fascicles (bundles of axons). The presence of extracellular ephrins can drive axons to fasciculate (join together) with one another in order to avoid the inhibitory signal. Such a role appears to be played by human ephrin-A5 (AL-1). Cortical neurons express the human ephrin-A5 receptor EphA5 (Rek7) and will fasciculate with one another when plated on astrocytes expressing human ephrin-A5. However, if this interaction is blocked, the axons will defasciculate (separate), suggesting that human ephrin-A5 normally makes the astrocytes a less favourable substrate and thus drives fasciculation. Therefore activation of the Eph receptors' kinase activity can either inhibit axonal extension or regulate fasciculation to determine axonal pathway choice.

Extracellular matrix and cell adhesion molecules

A variety of extracellular matrix (ECM) molecules and cell adhesion molecules (CAMs) have been implicated in axon guidance, primarily on the basis of their ability to promote outgrowth *in vitro* and of their expression patterns *in vivo*.

The major role for the ECM molecules, such as laminin, collagen and fibronectin, appears to be the provision of a permissive substrate for axonal growth. This substrate can be modulated, e.g. by the addition of proteoglycans that convert it into a non-permissive surface. Furthermore, the wide variety of ECM molecule isoforms suggests the potential to provide highly variable neuron outgrowth activities. However, conclusive evidence for a specific role for the ECM in guiding axons *in vivo* is lacking [17]. A key regulatory role for the ECM may come from an ability to present signalling molecules or the stabilization of molecular gradients.

The CAMs include the cadherins, integrins and members of the immunoglobulin superfamily [17]. The N-cadherins and integrins display a broad pattern of neural and non-neural expression, and their primary function appears to be in general adhesive mechanisms. However, mutations in *Drosophila* N-cadherin do produce growth-cone migration defects, suggesting that specific cadherins may well be critical for axon pathfinding [18].

The large group of immunoglobulin CAMs (IgCAMs) each contain tandemly arranged immunoglobulin domains that are extracellular (Figure 2). The IgCAMs promote neuron outgrowth *in vitro* and are able to bind one another homophilically (interaction between molecules of the same type) and/or heterophilically (interaction between molecules of different types). Many of the IgCAMs are expressed specifically along particular neural pathways and have been implicated in axon pathfinding through a role in promoting selective axon fasciculation, a process whereby axons choose to join (or leave) a pre-formed pathway. For example, transient axonal glycoprotein-1 (TAG-1) is expressed by rat spinal cord axons as they extend ventrally; however, when they reach the floor plate and turn longitudinally, TAG-1 is down-regulated and L1 expression is initiated. This switch correlates with the change in trajectory and an onset of fasciculation.

Genetic analysis in *Drosophila* has revealed a critical role in fasciculation for fasciclin II, the fly homologue of N-CAM (neural CAM). Overexpression of fasciclin II on motor axon surfaces prevents the selective defasciculation that allows individual axons to leave the common motoneuron pathway and enter their target area [1]. Moderation of N-CAM-mediated fasciculation also occurs in the chick, where it is regulated by polysialic acid. Enzymic removal of polysialic acid results in increased fasciculation and projection errors. Simultaneous application of antibodies to L1 (neuron–glia cell adhesion molecule; Ng-CAM) will overcome the effects of the enzyme treatment, suggesting that polysialic acid inhibits fasciculation provided by a heterophilic interaction between N-CAM and L1 [19].

Guidance molecules at intermediate targets

The use of intermediate targets by neurons as a mechanism to aid their migration across long distances raises a fascinating problem. The axons must reorient on reaching the intermediate target in order to continue on towards their final target, i.e. a favourable cue must now appear less preferable. This may be achieved by a neighbouring substrate providing greater outgrowth-stimulating properties to redirect the growth cone. An alternative mechanism is the provision by the intermediate target of a signal that modulates the receptors present on the growth cone, or a local cue that overrides a previously strong attractant. It appears that these latter types of signal may exist at the midline of the nervous system. Midline cells produce netrin, which attracts axons; this attractant needs to be overcome in order that axons can leave the midline. This may occur via the local production of a negative signal which axons only perceive once they reach the midline. Evidence for such a signal regulating axon pathway choice comes from the chick and *Drosophila*.

The IgCAMs axonin-1 and NgCAM-related cell adhesion molecule (Nr-CAM) regulate pathway choice by chick commissural axons at the floor plate. In the chick, axonin-1 (chick TAG-1) is expressed on the commissural axons before and after they cross the floor plate, while Nr-CAM is expressed at the floor plate. Interference with axonin-1 or Nr-CAM function causes defasciculation and a failure of the axons to cross the floor plate; instead they turn on their own side. This suggests that a heterophilic interaction between axonin-1 on the axons and Nr-CAM on floor-plate cells is necessary for axons to cross the floor plate and make appropriate pathway decisions [20]. Thus a negative signal may exist at the midline that is normally masked by the Nr-CAM–axonin-1 interaction. Interaction with the midline may induce a change in the repertoire of axonal surface receptors, unmasking this negative activity to allow axons to extend away from the midline.

In *Drosophila* the midline cells express netrin and also commissureless (comm). Comm appears to act to mask a negative guidance cue present at the midline that is normally recognized by the receptor protein roundabout (robo). In the absence of comm the commissural axons that normally link the two sides of the nervous system fail to cross the midline. In *robo* mutants, longitudinal axons (which never normally cross) cross and re-cross the midline, since the negative cue is no longer recognized. Normal crossing is achieved by comm down-regulation of robo on commissural axons to allow them across the midline [21]. This complex interplay between guidance signals may be necessary for accurate migration via intermediate targets.

Signal transduction of axon guidance signals

Extension of a neuron occurs by the localized assembly of actin filaments at the membrane on the leading edge of the growth cone, driving it forwards. The growth cone makes steering decisions by local changes to its cytoskeleton, so directing extension in an appropriate direction. Therefore the ultimate result of the various signalling molecules has to be a reorientation of the motile machinery of the growth cone. This is the same whether the actual signal is produced at a distance or locally, as it must eventually impinge on the growth-cone surface and the information be transduced into the cell. Yet how the external signals activate their receptors and transduce this information is presently not fully understood. Clearly, levels of tyrosine phosphorylation must play a part, as judged by the role played by the Eph receptor tyrosine kinases, the observation that the fibroblast growth factor receptor tyrosine kinase is activated *in vitro* to convey the signal for axonal outgrowth stimulated by CAMs [22], and the high levels of the non-receptor tyrosine kinases Src, Fyn and Yes present in growth cones. Furthermore a role for phosphatases in axon guidance has also been characterized in *Drosophila*, where four receptor protein tyrosine phosphatases have been identified in the developing central nervous system (DLAR, DPTP69D, DPTP99A and DPTP10D). It is also intriguing to note that inhibition of cAMP-dependent protein kinase A in the growth cone converts its response to a gradient of netrin from one of attraction to one of repulsion [23]. However, the identities of the phosphorylated substrates in all cases are uncharacterized.

It seems likely that the extracellular signals are transduced via the regulation of the Rho family of small GTP-binding proteins (RhoA, Rac1 and Cdc42). RhoA, Rac1 and Cdc42 differentially modulate the actin cytoskeleton, and interference with their function can cause defects in axon outgrowth and dendrite formation in *Drosophila* and Purkinje cells [24]. Dominant-negative Rac1 can also inhibit the collapsin 1-induced collapse of DRG growth cones, while constitutively active Rac1 increases the proportion of collapsed growth cones [25]. However, the nature of the adaptor molecules that link the receptors on the surface of the growth cone to the GTP-binding proteins are presently unknown. One candidate could be the Src homology 2 (SH2)/SH3 adaptor protein Nck that is the vertebrate homologue of *Drosophila* dreadlocks, a protein required for photoreceptor axon guidance. Nck interacts physically with the Wiskott–Aldrich syndrome protein (WASP), which is capable of binding the GTP-bound form of Cdc42 and can induce actin polymerization. Therefore these intracellular molecules have the potential to link the events at the growth-cone surface to the rearrangement of cytoskeletal structures.

Future perspectives

The last 10 years have seen a large increase in our understanding of the molecular mechanisms that establish neural connectivity. This has come from the use of improved *in vitro* assays and the investigation of function *in vivo* through genetic analysis and experimental perturbation. The discovery of diverse but evolutionarily conserved ligand and receptor families that provide multiple functional activities to direct axons to their targets has generated much excitement within the field. How the different signals are integrated during development and the mechanisms by which these signals are converted so as to direct the growth cone remain to be elucidated. Are there further families of signalling molecules and receptors awaiting discovery? How does a growth cone receiving a myriad of positive and negative cues always make the correct decision about where to extend? What is the nature of the intracellular signalling molecules that transduce these signals? How can we extend and use our knowledge of developmental events to help encourage the growth of injured or diseased neural tissue? With the molecules discovered thus far in hand, the way is open to identify the molecules that convert the extracellular signals in order to control precisely the dynamics of the growth cone. Molecular-interaction screens such as the yeast two-hybrid technique will allow us to begin to characterize further members of these signalling pathways. Genetic screens in a number of organisms will enable the discovery of new genes that regulate the function of identified molecules and reveal new signalling molecules. Through these techniques, it is likely that the next few years will see the continued characterization of signalling molecules, the mechanisms that regulate their expression and how growth cones respond to their extracellular cues. Hopefully we will soon be in a position to fully describe the molecular basis of axon pathfinding from initial outgrowth to final target selection.

Summary

- *Diffusible and substrate-bound cues can guide axonal pathway choice via attractive and repulsive signals.*
- *A number of families of signalling molecules have been identified, including netrins and their receptors, semaphorins, neuropilins, Eph receptor tyrosine kinases, ephrins and CAMs.*
- *Many of these signalling molecules can have a dual role, functioning either as attractants or as repellents.*
- *Direction of growth cone extension requires reorganization of the cytoskeleton, which may be directed by the Rho family of GTPases.*

My thanks to D. Hartley and the reviewers for comments on the manuscript. I also thank the MRC for a Senior Research Fellowship, and the MRC and BBSRC for supporting research in my laboratory.

References

1. Tessier-Lavigne, M. & Goodman, C.S. (1996) The molecular biology of axon guidance. *Science* **274**, 1123–1133
2. Kennedy T.E., Serafini, T., de la Torre, J.R. & Tessier-Lavigne, M. (1994) Netrins are diffusible chemotropic factors for commissural axons in the embryonic spinal cord. *Cell* **78**, 425–435

3.   Serafini, T., Colamarino, S.A., Leonardo, E.D., Wang, H., Beddington, R., Skarnes, W.C. & Tessier-Lavigne, M. (1996) Netrin-1 is required for commissural axon guidance in the developing vertebrate nervous system. *Cell* **87**, 1001–1014

4.   Wadsworth, W.G., Bhatt, H. & Hedgecock, E.M. (1996) Neuroglia and pioneer neurons express UNC-6 to provide global and local netrin cues for guiding migrations in *C. elegans. Neuron* **16**, 35–46

5.   Colamarino, S.A. & Tessier-Lavigne, M. (1995) The axonal chemoattractant netrin-1 is also a chemorepellent for trochlear motor axons. *Cell* **81**, 621–629

6.   Fazeli, A., Dickinson, S.L., Hermiston, M.L.,Tighe, R.V., Steen, R.G., Small, C.G., Stoeckli, E.T., Keino-Masu, K., Masu, M., Rayburn, H., et al. (1997) Phenotype of mice lacking functional *Deleted in colorectal cancer (Dcc)* gene. *Nature (London)* **386**, 796–804

7.   Kolodziej, P.A., Timpe, L.C., Mitchell, K.J., Fried, S.R., Goodman, C.S., Jan, L.Y. & Jan, Y.N. (1996) *frazzled* encodes a *Drosophila* member of the DCC immunoglobulin subfamily and is required for CNS and motor axon guidance. *Cell* **87**, 197–204

8.   Leonardo, E.D., Hinck, L., Masu, M., Keino-Masu, K., Ackerman, S.L. & Tessier-Lavigne, M. (1997) Vertebrate homologues of *C. elegans* UNC-5 are candidate netrin receptors. *Nature (London)* **386**, 833–838

9.   Luo, Y., Raible, D. & Raper, J.A. (1993) Collapsin: A protein in brain that induces the collapse and paralysis of neuronal growth cones. *Cell* **75**, 217–227

10.   Messersmith, E.K., Leonardo, E.D., Shatz, C.J., Tessier-Lavigne, M., Goodman, C.S. & Kolodkin, A.L. (1995) Semaphorin III can function as a selective chemorepellent to pattern sensory projections in the spinal cord. *Neuron* **14**, 949–959

11.   Koppel, A.M., Feiner, L., Kobayashi, H. & Raper, J.A. (1997) A 70 amino acid region within the semaphorin domain activates specific cellular response of semaphorin family members. *Neuron* **19**, 531–537

12.   Taniguchi, M., Yuasa, S., Fujisawa, H., Naruse, I., Saga, S., Mishina, M. & Yagi, T. (1997) Disruption of *semaphorinIII/D* gene causes severe abnormality in peripheral nerve projection. *Neuron* **19**, 519–530

13.   Kolodkin, A.L., Levengood, D.V., Rowe, E.G., Tai, Y.T., Giger, R.J. & Ginty, D.D. (1997) Neuropilin is a semaphorin III receptor. *Cell* **90**, 753–762

14.   Chen, H., Chédotal, A., He, Z., Goodman, C.S. & Tessier-Lavigne, M. (1997) Neuropilin-2, a novel member of the neuropilin family, is a high affinity receptor for the semaphorins Sema E and Sema IV but not Sema III. *Neuron* **19**, 547–559

15.   Drescher, U., Bonhoeffer, F. & Müller, B.K. (1997) The Eph family in retinal axon guidance. *Curr. Opin. Neurobiol.* **7**, 75–80

16.   Monschau, B., Kremoser, C., Ohta, K., Tanaka, H., Kaneko, T., Yamada, T., Handwerker, C., Hornderger, M.R., Löschinger, J., Pasquale, E.B., et al. (1997) Shared and distinct functions of RAGS and ELF-1 in guiding retinal axons. *EMBO J.* **16**, 1258–1267

17.   Bixby, J.L. & Harris, W.A. (1991) Molecular mechanisms of axon growth and guidance. *Annu. Rev. Cell Biol.* **7**, 117–159

18.   Iwai, Y., Usui, T., Hirano, S., Steward, R., Takeichi, M. & Uemura, T. (1997) Axon patterning requires DN-cadherin, a novel neuronal adhesion receptor, in the *Drosophila* embryonic CNS. *Neuron* **19**, 77–89

19   Tang, J., Rutishauser, U. & Landmesser, L. (1994) Polysialic acid regulates growth cone behavior during sorting of motor axons in the plexus region. *Neuron* **13**, 405–414

20.   Stoeckli, E.T. & Landmesser, L.T (1995) Axonin-1, Nr-CAM and Ng-CAM play different roles in the *in vivo* guidance of chick commissural neurons. *Neuron* **14**, 1165–1179

21.   Kidd, T., Russell, C., Goodman, C.S. & Tear, G. (1998) Dosage sensitive and complementary functions of roundabout and commissureless control axon crossing of the CNS midline. *Neuron* **20**, 25–33

22.   Doherty, P. & Walsh, F.S. (1996) CAM–FGF receptor interactions: A model for axonal growth. *Mol. Cell. Neurosci.* **8**, 99–111

23. Ming, G.-L., Song, H.-J., Berninger, B., Holt, C.E, Tessier-Lavigne, M. & Poo, M.-M. (1997) cAMP-dependent growth cone guidance by netrin-1. *Neuron* **19**, 1225–1235

24. Luo, L., Jan, L.Y. & Jan, Y.N. (1997) Rho family small GTP-binding proteins in growth cone signalling. *Curr. Opin. Neurobiol.* **7**, 81–86

25. Jin, Z. & Strittmatter, S.M. (1997) Rac1 mediates collapsin-1-induced growth cone collapse. *J. Neurosci.* **17**, 6256–6263

2

Understanding neurotransmitter receptors: molecular biology-based strategies

Mark Wheatley

School of Biochemistry, University of Birmingham, Edgbaston, Birmingham B15 2TT, U.K.

Introduction

For a cell to be responsive to a neurotransmitter, it must express an appropriate receptor protein on its surface. These receptors have to perform the two basic functions of binding the natural agonist when it is present in the extracellular milieu and generating a signal inside the cell. Although various families of receptor proteins have been described, neurotransmitter receptors can be classified as either G-protein-coupled receptors (GPCRs) or ligand-gated ion channels (LGICs), as shown in Figure 1. Receptors within each family share a characteristic architecture. GPCRs consist of one polypeptide composed of seven hydrophobic transmembrane (TM) domains connected by extracellular and intracellular loops (Figure 1c), whereas LGICs are oligomeric and possess an integral ion channel (Figure 1d). Both classes of receptor are fundamentally important to the activity of individual neurons and also underlie complex phenomena such as memory, emotion and intellect. As such, the structure and function of neurotransmitter receptors is of interest not just to basic neuroscientists but also to the pharmaceutical industry, as they represent potential targets for therapeutic intervention. Consequently, this has been an area of sustained investigation. The application of molecular biology techniques has greatly increased our understanding of neurotransmitter receptor structure and

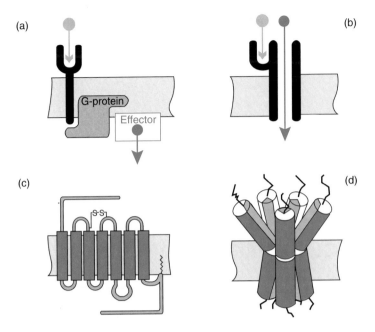

Figure 1. Signal transduction by neurotransmitter receptors
Schematic representation of (a) a GPCR and (b) a LGIC (each shown in black). In each case the light blue arrow represents neurotransmitter binding, and the dark blue arrow represents generation of an intracellular signal. (c) Schematic diagram of a GPCR, consisting of a single polypeptide with seven transmembrane domains (dark blue), indicating the location of the conserved disulphide bond and palmitoylated cysteine(s) (zig-zag line). In reality, the transmembrane domains are arranged like the staves of a barrel, as illustrated in Figure 3. (d) Schematic diagram of a LGIC, showing how each of the individual subunits of a pentameric receptor contributes to the formation of the ion channel.

function, and recent progress will be reviewed. Due to limitations of space, this article will focus predominantly on GPCRs.

How many receptors?

It might be thought reasonable that for every neurotransmitter there would be a corresponding receptor protein. Consequently, the number of receptors would be equivalent to the number of neurotransmitters. However, as long ago as 1914, Sir Henry Dale established that multiple receptors exist for the classical neurotransmitter acetylcholine. He demonstrated that physiological responses to acetylcholine could be mediated by either muscarinic receptors (activated by muscarine; blocked by atropine) or nicotinic receptors (activated by nicotine; blocked by curare). The natural agonist, acetylcholine, was non-selective and activated both receptor subtypes. This set the norm for receptor classification for the next 70 years, in that demonstrating the existence of receptor subtypes was dependent on the discovery of subtype-selective antagonists or agonists. Muscarinic acetylcholine receptors (mAChRs) were themselves subdivided into M_1 and M_2 receptors on the basis of different pharmacological profiles to selective ligands. In the early 1980s, this classification was extended to M_1, M_2 and M_3 receptors on the basis of differential affinity for certain antagonists, notably pirenzepine, AF-DX 116 and hexahydrosiladifenidol. Pirenzepine is selective for M_1 receptors (expressed by neuronal tissue) and is used clinically in the treatment of peptic ulcer disease by inhibiting vagally stimulated gastric acid secretion.

There are inherent problems, however, associated with the use of pharmacological profiles to define neurotransmitter receptor subtypes. For example, the discovery of very selective ligands, which are capable of revealing the subtle differences in receptor architecture, is entirely capricious, and differences in primary sequence (isoreceptors) may not necessarily generate pharmacological differences. In addition, pharmacological differences may not be due to receptor subtypes but may reflect differences in cell membrane composition or variation in post-translational modifications (glycosylation, phosphorylation or acylation). A further complication for GPCRs is that the affinity of ligands, particularly agonists, is influenced by the receptor:G-protein coupling state. Consequently, the unambiguous demonstration of receptor subtypes depends on primary sequence information. This was achieved for mAChRs in 1986, when Professor Shosaku Numa and co-workers cloned M_1 and M_2 receptors. These corresponded to the pharmacologically defined M_1 and M_2 receptors respectively [1]. At this time, cloning of neurotransmitter receptors was a major undertaking of almost Herculean proportions, as it required: (i) solubilization of active receptor; (ii) approx. 13 000-fold purification of the receptor protein by affinity chromatography; (iii) generation of proteolytic fragments; (iv) purification of receptor-derived peptides; (v) peptide sequencing; (vi) synthesis of degenerate oligonucleotide probes based on the amino acid sequence data; and (vii) screening of cDNA libraries for positive clones. Once clones encoding neurotransmitter receptors became available, sequence identity could be exploited to clone closely related receptors from different tissues or species. Employing less stringent screening conditions allowed the cloning of receptor subtypes or even entirely different neurotransmitter receptors that share structural identity. For example, the first dopamine receptor cDNA was cloned from rat using the hamster β_2-adrenergic receptor as the probe. Aligning sequences of different cloned GPCRs revealed regions of sequence identity. Oligonucleotide probes corresponding to these domains were used to clone the M_3 mAChR together with the hitherto unknown M_4 and M_5 subtypes. Consequently, whereas pharmacological approaches defined three mAChR subtypes, molecular biological techniques revealed the existence of five [1].

The proliferation of known receptor subtypes identified by homology cloning was not unique to mAChRs. For example, there are five dopamine receptors and fourteen 5-hydroxytryptamine (5-HT) receptors! Consequently, cloning neurotransmitter receptors has presented new therapeutic targets to the pharmaceutical industry and raises the prospect of improved drug selectivity. In addition, the use of specific, or degenerate, primers to amplify sequences by PCR has resulted in a proliferation of GPCR cDNAs. An alternative strategy was developed for cloning the tachykinin NK_2 receptor (NK_2R), which binds the neuropeptide substance K, and the technique has been quite widely used subsequently for cloning other neurotransmitter receptors. A cDNA library was transcribed *in vitro* and the mRNA microinjected into *Xenopus* oocytes (see below). Substance K-evoked electrophysiological responses were obtained if full-length NK_2R mRNA was present in the mixture of mRNAs injected. Following stepwise fractionation of the library, a single functional cDNA was isolated by this 'expression screening' [2].

The result of all these various cloning strategies is that sequence databases currently contain the sequences of over 800 GPCRs from a variety of eukaryotes. Although GPCRs exhibit key structural features (see later), sequence identity does not necessarily mean that the cDNA obtained encodes a physiologically functional receptor. The recent explosion in genomic information and bioinformatic analysis of databases has generated over 100 putative GPCR clones for which there are no known agonists or physiological roles ('orphan receptors'). Occasionally, these orphan receptors are later identified, e.g. RDC4 encodes a 5-HT_{1D} receptor [3].

Although sequence information has expanded our knowledge of the plethora of receptors, it is not possible to classify subtypes on the basis of sequence alone, because the same receptor subtype expressed by a different species may have a different primary sequence. For example, the NK_1Rs from rat and human have 22 divergent residues out of 407, and the NK_2R expressed by the rat is eight residues shorter than the human homologue. Usually species-specific differences in sequence occur in domains that are unimportant for normal receptor function, as there is no evolutionary pressure to preserve structural motifs. However, the proliferation of synthetic analogues of naturally occurring neurotransmitters has resulted in some species differences in binding or signalling being manifested; e.g. bovine, rat and human V_{1a}-vasopressin receptors have different affinities for the extensively used antagonist [d(CH$_5$),Tyr(Me)^2]AVP (where AVP is [Arg^8]vasopressin) [4].

The first GPCRs cloned (β_2-adrenergic receptor and mAChRs) lacked introns, but subsequently it has been demonstrated that many GPCR genes do possess introns. This has resulted in splice variants adding a further layer of receptor heterogeneity. For example, the D_2-dopamine receptor exists in short (D_{2S}) and long (D_{2L}) forms. D_{2L} has an alternative spliced exon that generates a 29-amino-acid insert in the third cytoplasmic loop between residues 241 and 242 of D_{2S}. D_{2L} predominates, but there are only slight pharmacological differences between the two forms and these are restricted to the benzamide derivative drugs. However, it seems that inhibition of stimulated adenylate cyclase by D_{2L}, but not by D_{2S}, has an absolute requirement for a specific G-protein, $G_i\alpha2$ [5].

LGICs, such as ionotropic glutamate receptors (GluRs) and GABA$_A$ receptors (where GABA is γ-aminobutyric acid), are multi-subunit complexes for which many different subunits have been cloned to date. For example, there are six α, four β, four γ, one δ, one ϵ and three ρ subunits of the $GABA_A$ receptor, and that does not include splice variants. It has been established that different subunit composition confers different properties on the resulting receptor. As many of these receptors have a pentameric quaternary structure, this gives rise to a vast array of functional heterogeneity.

In addition to splice variants, receptor differences can be generated by mRNA editing [6]. AMPA (α-amino-3-hydroxy-5-methylisoxazolepropionate) and kainate receptors are LGICs and are subtypes of GluRs. The electrophysiological properties of these channels can be dramatically altered by editing of channel subunit mRNAs at several sites. Receptor mRNA editing capacity is important for normal brain function, since transgenic mice that were incompetent at editing the GluR-B subunit mRNA developed epilepsy and died soon after birth. Editing has now been reported for a GPCR; the 5-HT_{2C} receptor can have different residues at three positions in the second intracellular loop and this affects the efficiency of stimulation of its effector phosphoinositidase C, suggesting that mRNA editing regulates 5-HT signalling [7]. Research over the next few years will establish if this mechanism of regulating neurotransmitter receptor function is widespread.

Studies on human populations have revealed that some neurotransmitter receptors exhibit polymorphic variation. For example, there are at least six different polymorphisms of the β_2-adrenergic receptor, some of which affect receptor function. The N-terminus variants $Arg^{16} \rightarrow Gly$ and $Gln^{27} \rightarrow Glu$ differ in their down-regulation following exposure to agonist, whereas $Thr^{164} \rightarrow Ile$ variants exhibit different ligand binding and effector coupling [8]. Interestingly, the nocturnal form of asthma has been associated with the enhanced down-regulation shown by the Gly^{16} β_2-adrenergic receptor variant.

Approaches to determining the structures and physiological functions of native neurotransmitter receptors

Insights from the distribution and/or disruption of receptor expression
Pharmacological characterization of the individual receptor subtypes and detailed analysis of receptor structure/function at the molecular level were difficult in the past. This was because the specificity of the tools available was not usually sufficient to overcome the extreme complexity of the tissue. For example, we now know that many of the antagonists utilized to localize receptor distribution by autoradiography, or to characterize receptors by blocking their activation, were actually binding to several different receptor subtypes. Consequently, instead of using pharmacological probes such as radioligands to localize receptors, recent investigations have employed DNA probes. Detailed analysis of isoreceptor sequences enables cDNA probes to be constructed which are subtype-specific. Cellular distribution can then be

addressed by hybridization of the probes to the receptor mRNA using Northern blot analysis and/or *in situ* hybridization. This strategy can be sufficiently stringent to establish whether mRNAs for different isoreceptors are present in the same cell. By selecting the probe sequence carefully, detection can be restricted to an individual subtype or can include all members of a neurotransmitter receptor family.

Receptor blockade by an antagonist or covalent modification of G-proteins by bacterial toxins have been widely used to establish physiological roles for individual neurotransmitters. An alternative strategy is to reduce, or ablate, expression of individual receptors or G-proteins using molecular biological approaches. Injection or transfection of antisense oligonucleotides or full-length antisense cDNA has been employed to knock out components of the neurotransmitter signalling apparatus [9]. The rate of degradation of the target protein is crucial (proteins with a long half-life may take days to be affected) and the technique can be compromised by RNA secondary structure interfering with hybridization or by the antisense probe cross-hybridizing with non-target mRNA. These various factors often result in antisense probes generating a knock-down (a reduction) rather than a knock-out (an elimination). Therefore the extent of the protein knock-out must be quantified directly in each case.

Homologous recombination between chromosomal DNA and introduced cDNA can be used to transfer a modification, or disruption, of a cloned gene into the genome of living mice. In this way 'knock-out' (KO) mice have been genetically engineered in order to study the physiological function of neurotransmitter receptors. Despite the extraordinary costs involved, an ever-increasing number of KO mice have been developed to elucidate the physiological roles of neurotransmitter receptors. For example, this strategy has established that mice lacking the mGluR4 subtype of metabotropic GluRs have impaired synaptic plasticity. In addition, KO mice establish unambiguously if responsiveness to a neurotransmitter in a tissue/region is conferred by a single gene. Thus pharmacological differences had led to the suggestion that neuronal and smooth muscle B_2-bradykinin receptors were different subtypes, but ablation of the B_2 receptor eliminated bradykinin responses in both the superior cervical ganglia and the uterus [10]. As the absence of a receptor subtype simplifies interpretation of data, KO mice lacking one isoreceptor can be very useful for studying the roles of related subtypes still expressed by the mice. A potential problem with KO mice is that ablation of a neurotransmitter receptor may lead to flawed development. Consequently, it is likely that KO mice in the future will have targeted gene disruption that can be regulated with respect to time and/or tissue.

Insights from the expression of cloned recombinant receptors

The pharmacological properties of a receptor can be influenced by many factors, including variability in receptor reserve and differences in the cellular context, such as type/level of G-proteins, effectors and membrane lipid.

These complexities can be circumvented by expressing different receptor subtypes in an identical cellular environment using recombinant expression in cultured cells. Furthermore, a homogeneous receptor population is obtained if the cultured cell does not express the receptor endogenously. Obviously, the appropriate post-translational processing, G-proteins, effector systems, kinases etc. must be present to enable the receptors to be functional and suitably regulated. It is for this reason that mammalian cells are the most commonly used for both stable and transient expression, but important results have also been obtained using *Xenopus* oocytes, yeasts and insect cells transfected with baculovirus.

Expression of recombinant receptors in these systems has revealed that a single GPCR can couple to multiple effectors and that the signalling pathway to which a GPCR couples is dependent on the level of expression and cell type [11]. Consequently, more is not always better when considering receptor expression, particularly if data obtained in recombinant systems are being extrapolated to effects in the whole animal. It has long been recognized that the efficacy of agonists is influenced by the abundance of receptors. It was not surprising, therefore, that overexpression of GPCRs in recombinant systems often resulted in elevated basal activity. This can be explained by the receptors existing in two interconvertible forms, inactive (R) and active (R^*), which are in equilibrium. Agonists stabilize the active form, altering the position of the equilibrium in favour of R^*, whereas antagonists do not differentiate between R and R^* and, therefore, lack intrinsic activity. Increasing the receptor density can inadvertently increase the abundance of R^*, due to the pre-existing $R \leftrightarrow R^*$ equilibrium, and generate an elevated basal activity. However, studies on recombinant expression systems have revolutionized receptor theory. It was demonstrated that, although some antagonists have null intrinsic activity, many antagonists reduce basal activity and, therefore, possess negative efficacy (Figure 2). It was proposed that antagonists with negative efficacy stabilize R and shift the equilibrium in favour of the inactive conformation (Figure 2) [12]. These ligands are termed inverse agonists, a term originally used with respect to $GABA_A$ receptors to describe a class of benzodiazepines which induced a reduced affinity for GABA. As for classical agonists, they can exhibit a range of intrinsic activities.

Observations of inverse agonism with GPCRs have generally been restricted to recombinant overexpression systems, raising doubts as to its relevance. However, the theory does translate to whole animals, as transgenic mice with cardiac-specific overexpression of β_2-adrenergic receptors exhibited a maximal baseline atrial rate and contractility in the absence of agonist, and the antagonist ICI-118,551 functioned as an inverse agonist [13]. Consequently, inverse agonists could have a unique therapeutic role in treating disease states resulting from overexpression of receptors or constitutively active receptors where mutation has shifted the equilibrium from R towards R^*. Known examples of the latter are few, but include the luteinizing hormone receptor and the thyrotropic

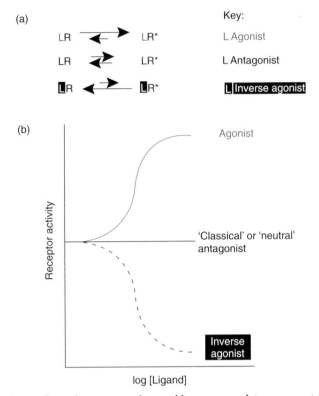

Figure 2. Effects of agonists, antagonists and inverse agonists on receptors
(a) The receptor exists in both an inactive state (R) and an active state (R*), which are in equilibrium. Binding of an agonist moves the equilibrium in favour of R*, in contrast with binding of an inverse agonist, which moves the equilibrium towards R. 'Classical' or 'neutral' antagonists do not perturb the equilibrium. (b) Agonists activate the receptor (blue line), inverse agonists inhibit the receptor (broken black line) and antagonists bind to the receptor but do not affect its activity (solid black line).

hormone receptor, the constitutive activities of which underlie familial male precocious puberty and hyperfunctioning thyroid adenomas, respectively.

Many neurotransmitter receptors signal by coupling to G_s or G_q to raise, respectively, cAMP or intracellular Ca^{2+}. Stable recombinant cell lines expressing reporter genes have been engineered in order to characterize receptor signalling and to facilitate high-throughput screening of putative ligands by pharmaceutical companies. For instance, a cAMP reporter cell line has been generated by stable expression in Chinese hamster ovary (CHO) cells of the firefly luciferase gene under the control of cAMP response elements. These cells generated a dose-dependent increase in luciferase expression in response to elevated cAMP [14]. In another example, receptor-mediated rises in intracellular Ca^{2+} were studied using cells stably co-transfected with the calcium-sensitive photoprotein, apoaequorin, and $G_{16}\alpha$. The G_{16} subunit was transfected following observations that it allowed a wide range of GPCRs to couple to phospholipase Cβ [15].

Receptor architecture and function

A primary objective in our understanding of how neurotransmitter receptors selectively bind ligands and generate intracellular signals is the identification of key functional domains/residues within the receptor protein. Two basic strategies have been employed: (i) replacement of individual residues by site-directed mutagenesis; and (ii) use of chimaeric receptors in which whole domains have been substituted by the corresponding domain from a different, but related, receptor. Engineered constructs are then characterized in recombinant expression systems. A potential problem with these techniques is that changes in function following structural modification may not necessarily indicate a direct role for a given domain or residue, but instead reflect changes in overall receptor conformation, membrane insertion, trafficking, protein half-life etc.

Ligand binding

The synergistic approaches of molecular biology and protein chemistry have resulted in an explosion of information aimed at defining the ligand binding site and receptor–G-protein contact sites for GPCRs. As space is limited, only the main themes arising out of the work from many laboratories are presented, and the reader is directed to reviews [1,16] as a source of primary references. The binding site for small ligands, such as the biogenic amines acetylcholine, dopamine, noradrenaline etc., is buried approx. 11–15 Å (1.1–1.5 nm) within the hydrophobic core of the GPCR. An Asp residue, positioned approximately one-third of the way down TM3, is conserved throughout this subfamily of GPCRs and provides a counter-ion to the charged amine head-group of both agonists and antagonists. A role for this Asp in ligand binding has been confirmed by site-directed mutagenesis [1,16] and by affinity labelling/radio-sequencing of the mAChR with [3H]propylbenzilylcholine mustard [17]. Other binding epitopes are provided by residues from TM5 and TM6. For example, the hydroxy groups of catecholamines and 5-HT hydrogen-bond with serine residue(s) located one turn apart in TM5. Residues at analogous positions to these serines have been implicated in agonist binding to mAChRs and histamine receptors. In fact, a series of conserved Ser, Thr and Tyr residues located on helices TM3–TM7 are in the same general plane as the conserved Asp in TM3. Mutation of any one of these decreases agonist, but not antagonist, binding, indicating subtle binding-site differences between the two classes of ligand. The occurrence of key residues, particularly the Asp in TM3, is diagnostic that a GPCR has a biogenic amine as the natural ligand, and this has aided subsequent identification of orphan receptors.

Size considerations alone dictate that the binding site for peptide neurotransmitters is not confined within the TM helical bundle, but must also

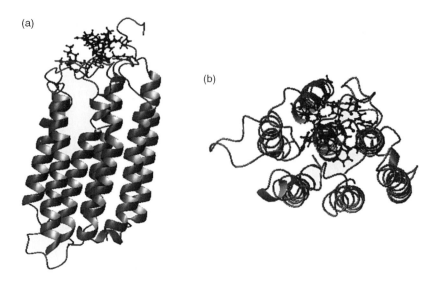

Figure 3. Computer model of the NK₁ receptor with substance P bound
The arrangement of the α-helical transmembrane domains of the NK1 receptor (a GPCR) can be seen with the ligand, substance P, shown in black. The receptor–ligand complex is shown (a) viewed through the plane of the membrane and (b) viewed from above. This figure was kindly provided by Paul R. Gouldson, Chris Higgs and Christopher A. Reynolds (University of Essex, Colchester, U.K.).

involve extracellular domains (Figure 3). Site-directed mutagenesis, deletion and chimaeric receptor constructs have indeed established that both the extracellular surface and the TM domains contribute to peptide ligand binding energy and ligand selectivity. For example, three residues in the N-terminus and three residues in the first extracellular loop are critical for high-affinity binding of neurokinin peptides to the NK₁ receptor. This contrasts with biogenic amine GPCRs, where extensive deletion of the extracellular domains was not detrimental to binding [16]. Mutation of individual residues has also indicated that the antagonist and agonist binding sites are not identical. This is particularly apparent for non-peptide antagonists, developed by the pharmaceutical industry for their superior stability relative to peptides. These are usually small molecules which bind to a pocket formed by the top of TM3–TM7, with no significant contribution from extracellular domains. Consequently, very few residues contribute to both the agonist and the non-peptide antagonist binding sites, despite the competitive binding observed. This implies that allostery or mutual exclusion effects, rather than overlapping intermolecular interactions, underlie competitive ligand binding between such compounds.

G-protein coupling and receptor activation
Molecular biological and biochemical strategies have established that multiple intracellular domains contribute to receptor–G-protein coupling. In particular, residues in the N- and C-terminal portions of the third intracellular loop (i3)

located near to the membrane have been implicated in the recognition and selection of G-proteins. Despite conservation of function, sequence identity is not displayed by these regions, even between related isoreceptors. However, homology is thought to exist at the level of secondary structure, as these domains are highly charged and are predicted by modelling programs to form amphipathic α-helices juxtaposed to the membrane. Synthetic peptides corresponding to GPCR intracellular loops can activate G-proteins *in vitro*, suggesting that the unoccupied receptor constrains the G-protein coupling domains in an inactive conformation (R). Agonist binding induces allosteric changes which relax this constraint, thereby facilitating receptor–G-protein interaction. Currently, these conformational changes associated with receptor activation are poorly understood. Mutation of the C-terminus of i3 has generated constitutively active receptors, indicating that an active receptor conformation (R^*) can be induced without agonists and suggesting a constraining function for this region of the i3 loop [18]. The conserved Asp-Arg-Tyr (DRY) motif at the base of TM3 is important for receptor–G-protein coupling. Molecular modelling and engineered constitutively active receptors have suggested that the Arg in this DRY motif acts as a switch during $R \leftrightarrow R^*$ conversion. Activation of receptor, by either agonist or mutation, causes the Arg side chain to move out of a polar pocket formed by TM helices 1, 3, 6 and 7, concomitantly favouring coupling of R to the G-protein [19].

GPCR dimerization

It has been suggested that TM1–TM5 and TM6–TM7 of GPCRs are independently folded. When these two regions were cloned in different vectors and coexpressed, a functional receptor was reconstituted, notwithstanding the absence of a covalent bond between TM5 and TM6. Co-expression of α_2/M_3 and M_3/α_2 chimaeric GPCRs reconstituted some functional mAChR activity indicative of intermolecular interaction, i.e. GPCR dimer formation [20]. Furthermore, an inactivating mutation in a GPCR could be 'rescued' by coexpression with the appropriate part of the wild-type receptor. This has culminated in the hypothesis that GPCRs undergo a rearrangement upon activation which involves 'domain swapping' [21]. What is not clear currently is whether dimerization is obligatory to function or merely coincidental. For example, dimerization of nicotinic AChRs occurs, but is not a prerequisite for acetylcholine-induced channel opening.

Future perspectives

The model of GPCR structure will continue to be refined. In particular, structural information on the loops is required so that their role in ligand recognition and G-protein coupling can be defined. Improved computer models will aid rational selective drug design by indicating differences between isoreceptors in ligand contact points. A major gap in our understanding is a precise molecular description of GPCR activation, including conformational changes and the effect of dimerization (if any). A high-resolution structure would result from crystallographic studies. Currently, it is very difficult to obtain high-quality crystals of membrane proteins in general, but enormous efforts are being made in this direction. Crystals of receptor when unoccupied and when binding agonist or inverse agonist will provide insight into receptor activation processes.

Summary

- *Neurotransmitter receptor proteins can be divided into GPCRs and ligand-gated ion channels. Members within each family utilize a similar signalling mechanism and are structurally related.*
- *Multiple receptor subtypes are expressed for individual neurotransmitters, generating orders of complexity that have been exploited for therapeutic intervention.*
- *GPCRs have been cloned and extensively studied by using a range of molecular biological strategies, particularly recombinant expression systems. These have located functional domains for ligand binding and G-protein coupling.*
- *A combination of molecular biology and computer modelling has increased our understanding of receptor activation and will aid drug design.*

Due to limitations of space, I must apologize to the many contributors to this field that I have not been able to cite in this article. I am grateful to Dr. David R. Poyner (University of Aston, Birmingham, U.K.) and Dr. Mary Keen (University of Birmingham) for their constructive comments. I thank the Wellcome Trust, the BBSRC and the MRC for supporting my research in the area of GPCRs.

References

1. Hulme, E.C., Birdsall, N.J.M. & Buckley, N.J. (1990) Muscarinic receptor subtypes. *Annu. Rev. Pharmacol. Toxicol.* **30**, 633–673
2. Levitan, E.S. (1988) Cloning of serotonin and substance K receptors by functional expression in frog oocytes. *Trends Neurosci.* **11**, 41–43
3. Zgombic, J.M., Weinshank, R.L., Macchi, M., Schecter, L.E., Branchek, T.A. & Hartig, P.R. (1991) Expression and pharmacological characterisation of a canine 5-hydroxytryptamine receptor subtype. *Mol. Pharmacol.* **40**, 1036–1042
4. Howl, J. & Wheatley, M. (1993) Hepatic vasopressin receptors (VPRs) exhibit species heterogeneity. *Comp. Biochem. Physiol.* **105C**, 247–250
5. Lui, Y.F., Jakobs, K.H., Rasenick, M.M. & Albert, P.R. (1994) G-Protein specificity in receptor–effector coupling – analysis of the roles of G_o and G_{i2} in GH_4C1 pituitary cells. *J. Biol. Chem.* **269**, 13880–13886
6. Bass, B.L. (1997) RNA editing and hypermutation by adenosine deamination. *Trends Biochem. Sci.* **22**, 157–162

7. Burns, C.M., Chu, H., Rueter, S.M., Hutchinson, L.K., Canton, H., Sanders-Bush, E. & Emeson, R.B. (1997) Regulation of serotonin-2C receptor G-protein coupling by RNA editing. *Nature* (*London*) **387**, 303–308

8. Liggett, S. (1995) Functional properties of human β_2-adrenergic receptor polymorphisms. *Notes Physiol. Sci.* **10**, 265–273

9. Wahlestedt, C. (1994) Antisense oligonucleotide strategies in neuropharmacology. *Trends Biochem. Sci.* **15**, 42–46

10. Borkowski, J.A., Ransom, R.W., Seabrook, G.R., Trumbauer, M., Chen, H., Hill, R.G., Strader, C.D. & Hess, J.F. (1995) Targeted disruption of a B_2 bradykinin receptor gene in mice eliminates bradykinin action in smooth muscle and neurons. *J. Biol. Chem.* **270**, 13706–13710

11. Albert, P.R. (1994) Heterologous expression of G-protein-linked receptors in pituitary and fibroblast cell-lines. *Vitam. Horm.* **48**, 59–109

12. Milligan, G., Bond, R.A. & Lee, M. (1995) Inverse agonism: pharmacological curiosity or potential therapeutic strategy? *Trends Pharmacol. Sci.* **16**, 10–13

13. Bond, R.A., Leff, P., Johnson, T.D., Milano, C.A., Rockman, H.A., McMinn, T.R., Apparsundaram, S., Hyek, M.F., Kenakin, T.P., Allen, L.F. & Lefkowitz, R.F. (1995) Physiological effects of inverse agonists in transgenic mice with myocardial overexpression of the β_2-adrenoceptor. *Nature* (*London*) **374**, 272–276

14. George, S.E., Bungay, P.J. & Naylor, L.H. (1997) Functional coupling of endogenous serotonin (5-HT$_{1B}$) and calcitonin (C1$_a$) receptors in CHO cells to a cyclic AMP-responsive luciferase reporter gene. *J. Neurochem.* **69**, 1278–1285

15. Stables, J., Green, A., Marshall, F., Fraser, N., Knight, E., Sautel, M., Milligan, G., Lee, M. & Rees, S. (1997) A bioluminescent assay for agonist activity at potentially any G-protein-coupled receptor. *Anal. Biochem.* **252**, 115–126

16. Strader, C.D., Fong, T.M., Tota, M.R. & Underwood, D. (1994) Structure and function of G-protein-coupled receptors. *Annu. Rev. Biochem.* **63**, 101–132

17. Curtis, C.A.M., Wheatley, M., Bansal, S., Birdsall, N.J.M., Eveleigh, P., Pedder, E.K., Poyner, D. & Hulme, E.C. (1989) Propylbenzilylcholine mustard labels an acidic residue in transmembrane helix three of the muscarinic receptor. *J. Biol. Chem.* **264**, 489–495

18. Kjelsberg, M.A., Cotecchia, S., Ostrowski, J., Caron, M.G. & Lefkowitz, R.J. (1992) Constitutive activation of the $\alpha 1_B$-adrenergic receptor by all amino acid substitutions at a single site – evidence for a region which constrains receptor activation. *J. Biol. Chem.* **267**, 1430–1433

19. Scheer, A., Fanelli, F., Costa, T., De Benedetti, P.G. & Cotecchia, S. (1996) Constitutively active mutants of the $\alpha 1_B$-adrenergic receptor: role of highly conserved polar amino acids in receptor activation. *EMBO J.* **15**, 3566–3578

20. Maggio, R., Vogel, Z. & Wess, J. (1993) Coexpression studies with mutant muscarinic/adrenergic receptors provide evidence for intermolecular crosstalk between G-protein-linked receptors. *Proc. Natl. Acad. Sci. U.S.A.* **90**, 3103–3107

21. Gouldson, P.R., Snell, C.R. & Reynolds, C.A. (1997) A new approach to docking in the β_2-adrenergic receptor which exploits the domain structure of G-protein-coupled receptors. *J. Med. Chem.* **40**, 3871–3886

3

Molecular analysis of neurotransmitter release

Giampietro Schiavo[*1] and Gudrun Stenbeck†

Molecular NeuroPathobiology Laboratory, Imperial Cancer Research Fund, 44 Lincoln's Inn Fields, London WC2A 3PX, U.K., and †Bone and Mineral Centre, University College London, 5 University Street, London WC1E 6JJ, U.K.

Introduction

A key step in the evolution of complex pluricellular organisms was the creation of a specialized pool of cells devoted to the processing and transduction of signals arriving from the external environment, leading to a suitable co-ordinated response of the organism. These specialized cells, a primitive functional nervous system, had to solve the problem of sustaining a reliable and efficient cross-talk with neighbouring cells. The winning strategy that evolved is based on the regulated and compartmentalized secretion of ligands, a mechanism at the basis of synaptic transmission in neurons. Two distinct pathways characterize neurotransmitter release: non-quantal molecular leakage and quantal release. While the former is still a matter of debate and, in general, accounts for a minority of total synaptic transmission, the latter mechanism has been clearly demonstrated and predicts that the generation of quanta is due to the fusion of neurotransmitter-containing small synaptic vesicles (SSVs) with specialized active zones at the presynaptic plasma membrane [1].

Although the complexity of neuroexocytosis is not yet fully understood, studies from several laboratories around the world have begun to provide a unitary view of the molecular steps preceding and following SSV fusion. As represented schematically in Figure 1, a SSV, after being loaded with neurotransmitter, is transported into close proximity with the presynaptic plasma membrane via interaction with the actin cytoskeleton. This transport phase is

[1]*To whom correspondence should be addressed.*

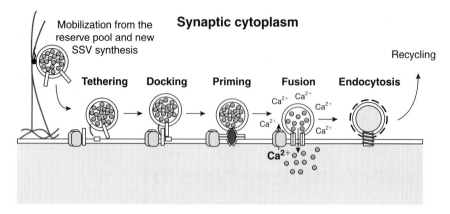

Figure 1. Late steps in the life cycle of an SSV
SSVs are mobilized from the reserve pool by the phosphorylation of synapsin, and are tied to the synaptic plasma membrane via specific protein–protein interactions. This tethering phase is followed by the functional docking of the SSV in the immediate vicinity of a Ca^{2+} channel and by priming, a reversible ATP-dependent phase. Exocytosis of the primed SSV is triggered by a very rapid increase in the intracellular Ca^{2+} concentration, generated by the opening of presynaptic voltage-gated Ca^{2+} channels following depolarization (fusion). The release of the SSV contents is via the opening of a fusion pore or complete fusion of the SSV with the presynaptic membrane. The empty SSV undergoes rapid endocytosis and is refilled with neurotransmitter by proton-driven neurotransmitter transporters.

followed by binding of the SSV to the active zone at the target membrane, in a process called tethering. This is followed by the functional docking of the SSV in the immediate vicinity of a Ca^{2+} channel and by an ATP-dependent phase, known as priming. Priming is a reversible process that could progress to the exocytic phase in the presence of a suitable concentration of Ca^{2+} (up to 200 μM), which is generated by the opening of presynaptic voltage-gated Ca^{2+} channels following depolarization. This rise in Ca^{2+} concentration causes, within 200–300 ms, the release of the SSV contents by means of the opening of a fusion pore or the complete fusion of the SSV with the presynaptic membrane [2,3]. After release, the empty SSVs undergo rapid re-uptake at the nerve terminal, followed by refilling with neurotransmitters by proton-driven neurotransmitter transporters.

In the last decade, the molecular analysis of neurotransmitter release has made the transition from a mere description of proteins expressed in the nervous system to a more sophisticated investigation of the functions that these proteins perform during the SSV life cycle. One of the first soluble proteins important for SSV docking and fusion to be identified was the so-called N-ethylmaleimide-sensitive factor (NSF), an ATPase that is involved in a variety of intracellular transport pathways [4]. NSF binds to its membrane receptors via another soluble protein, SNAP (soluble NSF accessory protein); hence the name SNAREs (SNAP receptors) for the membrane receptor proteins. As SNAREs are localized to different membrane compartments, the nomenclature v-SNARE for vesicular SNARE [i.e. VAMP (vesicle-associated membrane protein)] and t-SNARE for target membrane SNARE [i.e. SNAP-25 (synapto-somal-associated protein of 25 kDa)] was adopted. SNAREs have been shown to be important for membrane transport events other than neurotransmitter release, thus supporting the hypothesis, initially proposed by Rothman and colleagues [4], that each membrane-trafficking step is characterized by a unique SNARE pairing.

In this article, we review the synaptic proteins that have been demonstrated to be involved in SSV exocytosis and the roles of some protein–protein and lipid–protein interactions that are important in this process. The importance of rearrangement of the cytoskeleton for exocytosis will also be discussed, together with more recent models of SSV fusion and endocytosis.

SNARE proteins

Three proteins of the synaptic terminal are at the centre of increasing attention as key players in neurotransmitter release. These are the SSV-specific VAMP (also known as synaptobrevin), and syntaxin and SNAP-25, both of which are localized predominantly on the plasma membrane.

VAMP is a protein of 13 kDa that is localized to SSVs, dense-core granules and synaptic-like vesicles. Ten different isoforms have been identified on the basis of structural sequence similarity, but only three isoforms have been extensively characterized: VAMP-1, VAMP-2 and cellubrevin. VAMP isoforms are present in all vertebrate tissues, but their distribution varies [5]. Structurally, VAMP is composed of an N-terminal portion that is rich in proline residues and is divergent in different isoforms, a very conserved central portion that contains coiled-coil segments which are responsible for the pairing with the corresponding t-SNAREs (syntaxin and SNAP-25), and a single membrane-spanning domain. The majority of the protein mass is exposed to the cytoplasm, while only a short and poorly conserved portion is intravesicular (Figure 2). On SSVs, VAMP-2 is associated with synaptophysin, a major component of the SSV membrane, and with subunits of the V-ATPase. VAMP-2, but not VAMP-1 or cellubrevin, interacts with a prenylated Rab acceptor (termed PRA) via sequences that are present only in the proline-rich and the transmembrane domain of VAMP-2. The binding requirements for VAMP-2 and PRA are thus distinct from those characterizing the interaction of VAMP-2 with syntaxin and SNAP-25.

SNAP-25, originally described as the major palmitoylated protein in the central nervous system, is required for axonal growth during neuronal development and in nerve terminal plasticity in the mature nervous system. In the nervous system and in neuroendocrine cells, SNAP-25 is expressed in two isoforms which are developmentally regulated (SNAP-25A and B) [6]. Structurally, SNAP-25 lacks a classical transmembrane segment, and its membrane binding is mediated by the palmitoylation of cysteine residues located in the middle of the polypeptide chain and by additional, still unidentified, protein–lipid interactions. Palmitoylation is essential for correct targeting of the molecule to the plasma membrane. As with the other SNAREs, SNAP-25

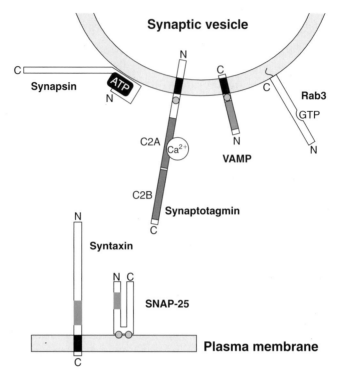

Figure 2. Schematic structures of some proteins involved in Ca²⁺-dependent exocytosis
Synapsins are characterized by an N-terminal globular domain which contains an ATP-binding site and which interacts with phospholipids, and by a C-terminal elongated tail region that binds actin. Palmitoylated cysteine residues (light blue circles) are present in synaptotagmin, VAMP and SNAP-25. Synaptotagmin also contains two C2 homology domains (C2A and C2B) that are involved in Ca²⁺-dependent and Ca²⁺-independent interactions with proteins and acidic phospholipids. VAMP, SNAP-25 and syntaxin are characterized by coiled-coil segments (dark blue) that are responsible for synaptic SNARE complex-formation. In VAMP-2, maximal binding to syntaxin requires almost all the central conserved region, while only the N-terminal portion of SNAP-25 seems to be essential for the interactions with syntaxin and VAMP. In syntaxin, the SNARE-interacting region is also involved in binding to α-SNAP and synaptotagmin. Rab3 is a GTP-binding protein of the Ras superfamily which is localized on SSVs and which interacts with the membrane bilayer through a C-terminal geranylgeranyl modification (~).

contains segments with the ability to form coiled-coil helices. SNAP-25 forms a stoichiometric complex with the putative Ca²⁺ sensor synaptotagmin I, and this interaction is believed to be important for a late step of the Ca²⁺-dependent phase of neurotransmitter release. In addition, SNAP-25 was demonstrated to interact in a Ca²⁺-dependent manner with Hrs-2, an ATPase having a negative regulatory effect on exocytosis.

Syntaxin is a typical type II membrane protein; its N-terminal portion is exposed to the cytosol, followed by a single transmembrane domain and few residues emerging into the intersynaptic space (Figure 2). Syntaxin is associated with N-, P- and Q-type Ca²⁺ channels in the active zones, where neurotransmitter release takes place, and it is also present on most of the neuronal cell membrane [5]. Syntaxin undergoes, together with SNAP-25, a recycling process in organelles indistinguishable from SSVs. It interacts in a Ca^{2+}-dependent manner with some isoforms of the SSV protein synaptotagmin. The minimal segment of syntaxin required for Ca^{2+}-dependent binding to synaptotagmin was localized to amino acids 220–266 (Figure 2). A more extended portion of this same region is responsible for the interaction with VAMP and SNAP-25, to assemble the synaptic SNARE complex, and for the interaction with α-SNAP. Deletional analysis confirmed that, in the other SNAREs also, regions with a high probability of forming coiled coils are essential for SNARE-complex formation.

Syntaxins constitute a large protein family, with more than a dozen different isoforms coded for by different genes or generated by alternative splicing, and a vast syntaxin polymorphism exists within the nervous tissue. Syntaxins are essential for neuronal development and survival. Several isoforms undergo a complex pattern of alternative splicing and expression control during long-term potentiation, suggesting that syntaxins are involved in synaptic plasticity. This differential expression could be important for a direct modulation of Ca^{2+} entry via selective interactions with specific Ca^{2+} channels, in addition to the formation of distinct SNARE complexes with different SNAP-25 and VAMP isoforms.

What is the precise role of the SNARE complex and its functional ligands, α-SNAP and NSF, in neurotransmitter release? The accumulation of vesicles in an *in vitro* intra-Golgi assay after NSF depletion, in the yeast NSF mutant *sec18* and in the corresponding *comatose* mutant of *Drosophila*, sustains the hypothesis that NSF could be involved in vesicle fusion with the target membrane [4]. The ability of NSF to bind via SNAPs to the SNARE complex and to disassemble this large 20 S particle via its ATPase activity further strengthens this proposal, and suggests that the disassembly of this large particle may represent the core of the membrane fusion mechanism. However, recent experiments have challenged this view and indicate that the action of NSF may be restricted to an earlier stage, the pre-docking and/or docking step [7]. In particular, NSF action could be restricted to the disassembly of SNARE complexes during the recycling of VAMP, SNAP-25 and syntaxin, allowing them to re-enter the docking and fusion cycle of the exocytic vesicles. Although compelling, these experiments cannot exclude a role for NSF in a post-docking stage of heterotypic membrane fusion or in modulation of the speed of neurotransmitter release [8].

The low-resolution structures of the SNARE complex and of 20 S particles have recently been determined using electron microscopic and rotary shadowing techniques [9]. The SNARE complex has a rod-shaped structure containing the SNARE proteins arranged in a parallel fashion (Figure 3). NSF (present as a hexamer) and α-SNAP occupy one end of the rod, and disappear when the particle is incubated in the presence of Mg^{2+}-ATP. The geometry of

(a)

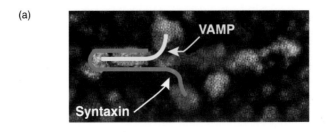

(b)

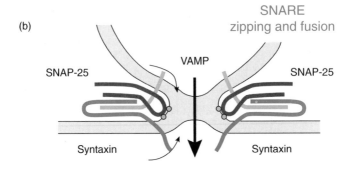

SNARE
zipping and fusion

(c)

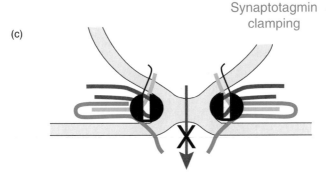

Synaptotagmin
clamping

Figure 3. Structure of the SNARE complex and mechanism of membrane fusion
(a) Low-resolution electron micrograph of the trimeric SNARE complex (modified from [9] with permission; © 1997 Cell Press). The rod-shaped SNARE complex contains VAMP-2 (light blue) and syntaxin (dark blue) arranged in a parallel fashion. (b) SNARE proteins are able to fuse two populations of artificial liposomes, one containing purified VAMP-2 and the other containing the complex between syntaxin (dark blue) and SNAP-25 (dark grey) respectively. The formation of the trimeric SNARE complex (SNARE zipping) is essential for Ca^{2+}-independent membrane fusion. (c) The Ca^{2+}-sensitivity of the fusion process could be provided by interaction of SNAREs with the Ca^{2+} sensor synaptotagmin (black), which might act by blocking the fusion reaction. This functional Ca^{2+} clamp could be removed as a result of the rise in intracellular Ca^{2+} concentration caused by the opening of Ca^{2+} channels.

the SNARE complex, the large amount of energy liberated by its formation and the observation that the cleavage of the SNAREs by clostridial neurotoxins leads to a block in SSV fusion and their subsequent accumulation on the plasma membrane suggest that the interaction of VAMP-2, SNAP-25 and syntaxin could be the basis of lipid bilayer fusion. Recent work from Rothman et al. [10] demonstrated that SNARE proteins are able to fuse artificial liposomes, in the absence of α-SNAP and NSF, following their interaction to form the trimeric complex. This complex, linking two membranes, named SNAREpins by the authors, has analogy with the viral fusogenic proteins containing hairpin-like structures and could be the basis of a general mechanism for protein-mediated lipid bilayer fusion.

Synaptotagmins as Ca^{2+} sensors at the synapse

Synaptotagmin is a type I membrane protein localized on SSVs that belongs to a growing family of proteins, with more than a dozen members present in nervous and non-nervous tissues [11]. Impairment of its function in neurons via gene knock-out approaches and by microinjection of peptides and antibodies strongly supports the possibility that synaptotagmin is the main Ca^{2+} sensor in the fast phase of neurotransmitter release [11]. Structurally, synaptotagmin is characterized by a large cytosolic domain, a single transmembrane region flanked on the cytoplasmic side by a cysteine-rich palmitoylated segment, and a short N-terminal portion in the lumen of the vesicles (Figure 2). The cytoplasmic portion of the protein is hydrophilic, contains two protein modules homologous to the C2 regulatory domain of protein kinase C and has a highly conserved C-terminus known to interact with members of the neurexin family. Moreover, synaptotagmin is able to interact with a variety of proteins and lipids via its C2 domains, both in a Ca^{2+}-dependent and in a Ca^{2+}-independent manner. Phosphatidylserine and syntaxin bind synaptotagmin via its C2A domain at Ca^{2+} concentrations similar to those observed at the synapse during neuroexocytosis. The C2B domain of synaptotagmin mediates the binding of SNAP-25, which is weakly dependent on Ca^{2+}, and the binding of β-SNAP, which is Ca^{2+}-insensitive. The C2B domain is also responsible for the Ca^{2+}-dependent hetero- and homo-dimerization of synaptotagmin. In fact, the presence of multiple synaptotagmin isoforms on a single SSV suggests that a combinatorial range of Ca^{2+} sensors could be created by heterodimerization of isoforms with different Ca^{2+} affinities. As a consequence, the probability of an SSV being released at a certain Ca^{2+} concentration is directly dependent on the repertoire of synaptotagmin isoforms present on its surface (Osborne, S.L., Herreros, J., Bastiaens, P.I. and Schiavo, G., unpublished work).

The range of synaptotagmin interactions with negatively charged lipids was recently expanded by the finding that the C2B domain binds phosphoinositides. In particular, PtdIns(3,4,5)P_3 binds to synaptotagmin in the absence of Ca^{2+}, whereas PtdIns(4,5)P_2 binds at micromolar Ca^{2+} concentrations. This equilibrium constitutes a Ca^{2+}-dependent switch that has the potential to localize the cytoplasmic portion of synaptotagmin to phosphoinositide-rich domains of the lipid bilayer. In addition, the interaction of phosphoinositides with synaptotagmin is the first direct linkage between a member of the protein exocytosis apparatus and these tightly regulated and essential lipid components of the membrane (see below).

Recent evidence suggests another important role for synaptotagmin in nerve-terminal physiology through its interaction with voltage-gated Ca^{2+} channels. N-type Ca^{2+} channels bind synaptotagmin directly via a cytosolic loop, and this binding fully restores the current amplitude and inactivation kinetics of these syntaxin-modulated channels. Immunoprecipitation experiments indicate that synaptotagmin also associates with P- and Q-type Ca^{2+} channels and that these complexes recruit the SNARE complex. At equilibrium, a large percentage of presynaptic Ca^{2+} channels is engaged in this association. Taken together, these results suggest that these proteins may constitute an isolated exocytic complex in which the Ca^{2+} channel interacts tightly with the SSV docking site.

Rab3 and neurotransmitter release

All trafficking steps throughout the secretory pathway are regulated by members of the Rab family of small GTP-binding proteins. Rab proteins do not contain a transmembrane region, but interact directly with the lipid bilayer through a C-terminal geranylgeranyl modification (Figure 2). Two members of this family (Rab3A and Rab3C) are suggested to play a role in regulated secretion, based on their exclusive expression in neurons and neuroendocrine cells. However, despite massive efforts to define the precise molecular action of Rabs in this process, the details of their action remain unknown, and have been assigned to the phase of docking and fusion of vesicles with the target membranes [11]. The function(s) of the Rab proteins is thought to be dependent on their GTPase activity as well as their effectors. The cycling of Rabs from the GDP- to the GTP-bound state determines their translocation to the target membranes. At least three types of activities are required for the Rab3 cycle: (1) a GTPase-activating factor triggering GTP hydrolysis; (2) a GDP-dissociation inhibitor promoting the removal of the Rab3–GDP from SSVs after exocytosis; and (3) a GDP-exchange protein catalysing GDP–GTP exchange. In this cycle, a pool of soluble Rab3 is maintained by the interaction with the GDP-dissociation inhibitor to shield the geranylgeranyl group [12].

Rab3A interacts *in vitro* with SNARE proteins and with the SNARE complex, but a direct effect on SNARE complex assembly has been clearly identified only in an early step of the secretory pathway in yeast. The lack of validation of this finding in other trafficking steps, and in neurotransmitter release in particular, is complicated by the absence of a severe phenotype in Rab3A-deficient mice. The only observed anomaly in hippocampal neurons lacking Rab3A is an increase in evoked quantal release after a single stimulation. This result was interpreted as an inhibition of multivesicular fusion mediated by Rab3A. Thus the function of Rab3A at the nerve terminal would be to restrict exocytosis to a single vesicle per releasing site [11], possibly through the interaction with a new Rab effector, termed Rim (Rab-interacting protein) in mammals. Rim binds only to GTP–Rab3 and not to GDP–Rab3, and is localized to presynaptic active zones in conventional synapses. Rim over-expression has a stimulatory effect on regulated exocytosis, possibly by reversing the inhibitory function of Rab3A.

Many putative Rab effectors have been described, but the lack of functional data makes speculation premature. The first Rab3A effector discovered was rabphilin 3A, which is specific for GTP–Rab3A. This soluble protein is associated with SSVs and contains two C2 homology domains and a zinc finger. Rabphilin 3A also interacts with cytoskeletal proteins and could be important for actin remodelling [11].

Phosphoinositide biosynthesis and turnover at the nerve terminal

Several lines of evidence support the pivotal role of phosphoinositides and the enzymes regulating their turnover in membrane-trafficking events. Three proteins involved in phosphoinositide biosynthesis were demonstrated to be essential for priming in Ca^{2+}-dependent exocytosis in neuroendocrine cells. The first of these factors is the phosphatidylinositol transfer protein, a soluble protein responsible for the mobilization of PtdIns and its correct presentation to lipid kinases. PtdIns transfer protein was found to be an essential cofactor in the maintenance of a primed ATP-dependent state in PC12 cells, together with PtdIns 4-phosphate 5-kinase, another soluble cytosolic protein. The activity of PtdIns 4-kinase, an integral membrane protein of secretory granules, is also necessary to sustain vesicle priming, thus strongly supporting a role for PtdIns(4,5)P_2 during the terminal phases of secretory granule exocytosis [13]. Although similar studies on SSVs are still lacking, the reversible inhibitory activity of inositol polyphosphates on neurotransmitter release in giant squid axons suggests that PtdIns(4,5)P_2 plays an essential role in both of these regulated exocytic events. The molecular basis of its action is still unclear. Direct involvement of PtdIns(4,5)P_2 in lipid bilayer fusion is unlikely, based on the high bivalent cation concentration needed to promote aggregation and fusion of PtdIns(4,5)P_2-containing liposomes. In addition, the high curvature caused by this phospholipid in the membrane bilayer would be expected to destabilize the vesicular fusion intermediates. Several proteins of the exocytic machinery interact with PtdIns(4,5)P_2, including CAPS (a novel Ca^{2+}-binding protein) [14] and synaptotagmins [15]. This latter binding is localized to the C2B domain and is blocked by InsP_6. Microinjection of a C2B-specific antibody into the giant squid axon reverses the InsP_6-evoked inhibition of neurotransmitter release, suggesting a direct action of InsP_6 on synaptotagmin–PtdIns(4,5)P_2 binding and a functional connection between phosphoinositides and synaptotagmin at the nerve terminal.

Cytoskeleton and exocytosis

Several lines of evidence suggest a central role for the cytoskeleton in the mobilization of SSVs and, more generally, the existence of a cross-talk between the cytoskeleton and members of the exocytosis machinery. At the synaptic bouton in mature synapses, the SSV distribution is determined by the interaction of synapsins with the cytoskeleton. The four members of this family of phosphoproteins (synapsins Ia, Ib, IIa and IIb) are generated by differential mRNA splicing from two distinct genes [16]. Synapsins have two domains: an N-terminal hydrophobic head region and a C-terminal tail region that is rich in polar residues (Figure 2). Crystallographic studies revealed that the N-terminus of synapsin shares structural similarities with a family of ATP-utilizing enzymes and binds ATP in a Ca^{2+}-dependent manner. The C-terminal region is responsible for the binding of synapsin to SSVs, an event which is regulated by phosphorylation. Synapsins are excellent substrates for cAMP-dependent protein kinase and Ca^{2+}/calmodulin-dependent protein kinase I. In the dephosphorylated state, synapsin I is able to bind actin filaments and to promote actin polymerization *in vitro*. These findings suggest that synapsins modulate the interaction of SSVs with the cytoskeleton at the nerve terminal [17]. The rapid phosphorylation of synapsins, triggered by the rise in Ca^{2+} concentration in the region proximal to the active site, disassembles the ternary complex formed by synapsin, SSV and actin, allowing the released SSV to enter the tethering phase of neurotransmitter release. This is supported by gene knockout experiments, which demonstrated that synapsins are essential for accelerating SSV availability during repetitive stimulation [18]. Synapsin I- or synapsin II-null mice are viable and fertile, but experience seizures with a frequency proportional to the number of mutant alleles. Synapsin II/I double knock-outs exhibit decreased post-tetanic potentiation and severe synaptic depression upon repetitive stimulation [18].

Another connection between the cytoskeleton and SSVs is provided by the molecular interaction of brain myosin V, a member of the family of unconventional myosins, and the binary complex formed between VAMP-2 and synaptophysin. Brain unconventional myosins are localized within the presynaptic terminals and are capable of generating mechanochemical force and moving actin filaments. In particular, myosin V appears to be involved in the transport of axoplasmic organelles. The interaction between myosin V and the synaptophysin–VAMP-2 complex is disrupted by Ca^{2+}, suggesting that a remodelling of this association occurs during the Ca^{2+}-dependent phase of neurotransmitter release [19]. The importance of this complex for Ca^{2+}-regulated exocytosis was confirmed in embryonic sea urchin cells by a microinjection approach that also revealed the involvement of members of the kinesin family.

Proteins involved in synaptic vesicle endocytosis

Synaptic vesicle endocytosis is strictly coupled to exocytosis. The most widely accepted model of synaptic vesicle recycling proposes that the SSV membrane is retrieved from the plasma membrane through both clathrin-mediated endocytosis [20] and a clathrin-independent process termed 'kiss and run' [21]. In the latter, the fusion of the vesicle with the synaptic plasma membrane is restricted to a fusion pore, with very limited, if any, intermixing between vesicle and plasma membrane components. The 'kiss and run' model can account for the high speed and specificity of membrane retrieval under physiological conditions. In clathrin-dependent recycling, the budding of the SSV from the plasma membrane is followed by excision that is dependent on dynamin. Dynamin binds GTP and oligomerizes into rings at the stalks of endocytic pits, closing the neck to release the recycling clathrin-coated SSV [20]. In addition, dynamin interacts with cytosolic factors essential for the coating as well as for the fission reaction. These are clathrin heavy and light chains, the clathrin adaptor complex AP2, AP180 and amphiphysins. Recent evidence indicates a crucial role for dynamin and amphiphysin in the recruitment of an inositol 5-phosphatase called synaptojanin [20]. The presence of synaptojanin on the recycling SSV strongly supports the hypothesis that, in addition to exocytosis, endocytosis is also tightly coupled with phosphoinositides, and offers a link between the cycle of phosphorylation/dephosphorylation of these lipids and SSV turnover.

Perspectives

The cloning of the major components of SSVs, together with genetic analysis in yeast, *Caenorhabditis elegans* and *Drosophila* and modern electrophysiological techniques, has provided us with invaluable insights into the molecular mechanism of neurotransmitter release. Despite these efforts, several points remain to be clarified. One of these is how synaptotagmin is able to confer Ca^{2+} sensitivity to the basic and Ca^{2+}-independent fusion mediated by VAMP-2, SNAP-25 and syntaxin. Experimental evidence suggests that synaptotagmin binds individual SNAREs both in a Ca^{2+}-dependent and a Ca^{2+}-independent manner, but the precise dynamics of the SNARE complex interaction with synaptotagmin in relation to the lipid membrane remain unclear. The experimental system used by Rothman and co-workers [10] offers the potential to investigate this important mechanism with purified components, and it could be extremely useful in verifying the hypothesis illustrated in Figure 3, which envisages synaptotagmin acting as a Ca^{2+}-dependent clamp. In this model, synaptotagmin would arrest the SNARE-induced fusion of the two lipid bilayers by stabilizing the fusion intermediate via a series of protein–protein and lipid–protein interactions, awaiting Ca^{2+} in order to progress to the fully competent fusion state.

A further level of complexity that remains largely unexplored is the large number of isoforms characterizing the protein machinery for neurotransmitter release. Although it is clear that the abundance of isoforms and splice variants is related to the fine regulation of the exocytic process in the central nervous system, their importance in relation to physiological processes, such as long-term potentiation and depression and synaptic plasticity, are presently unknown and deserve a much broader investigation.

In addition to the protein complexes described in the present review, a new large particle has been described recently in both yeast and mammals [22,23]. This large protein complex, termed exocyst, contains a new repertoire of proteins and is localized specifically at the site of vesicular fusion. In yeast, this particle localizes to sites of polarized exocytosis and its distribution is independent of the integrity of the actin cytoskeleton. In neuroendocrine cells, it is found in vesicle-rich processes of growing neurites and near sites of granule exocytosis. These features, which differ from those characterizing the synaptic SNARE proteins, makes the mammalian homologue of exocyst a very interesting candidate for the machinery controlling neurite elongation and for an alternative pathway for regulated exocytosis.

Summary

- *The synaptic vesicle cycle can now be subdivided into a series of defined steps. These are the tethering of the SSV at the active site of the presynaptic membrane, followed by docking and by an ATP-dependent phase, termed priming. Fusion of the primed SSV with the plasma membrane is triggered by Ca^{2+} entry through specific Ca^{2+} channels.*
- *All the steps in the SSV life cycle are regulated through a cascade of protein–protein and lipid–protein interactions.*
- *The SNARE proteins VAMP-2, SNAP-25 and syntaxin are essential for membrane fusion.*
- *Synaptotagmin is the major Ca^{2+} sensor for Ca^{2+}-regulated exocytosis at the synapse.*
- *Rab3A and Rab3C regulate SSV neurotransmitter release by restricting exocytosis to a single vesicle per releasing site.*
- *The mobilization and the availability of SSVs are regulated by interactions with the actin cytoskeleton.*
- *Specific phospholipids, such as phosphoinositides, are essential in order to sustain both exocytosis and endocytosis.*

We apologize to colleagues for not citing a number of relevant references owing to space limitations. We thank Dr. P.I. Hanson and Dr. J.E. Heuser for allowing the reproduction of the electron micrograph shown in Figure 3, and Dr. T. Iglesias and Dr. S. Tooze for critical reading of the manuscript. The work of G.St. is supported by the Wellcome Trust.

References

1. Angleson, J.K. & Betz, W.J. (1997) Monitoring secretion in real-time – capacitance, amperometry and fluorescence compared. *Trends Neurosci.* **20**, 281–287

2. Rahamimoff, R. & Fernandez, J.M. (1997) Pre- and postfusion regulation of transmitter release. *Neuron* **18**, 17–27

3. Palfrey, H.C. & Artalejo, C.R. (1998) Vesicle recycling revisited – rapid endocytosis may be the first step. *Neuroscience* **83**, 969–989

4. Rothman, J.E. (1994) Mechanisms of intracellular protein transport. *Nature (London)* **372**, 55–63

5. Bennett, M.K. & Scheller, R.H. (1994) A molecular description of synaptic vesicle membrane trafficking. *Annu. Rev. Biochem.* **63**, 63–100

6. Wilson, M.C., Mehta, P.P. & Hess, E.J. (1996) SNAP-25, enSNAREd in neurotransmission and regulation of behaviour. *Biochem. Soc. Trans.* **24**, 670–676

7. Mayer, A., Wickner, W. & Haas, A. (1996) Sec18p (NSF)-driven release of Sec17p (alpha-SNAP) can precede docking and fusion of yeast vacuoles. *Cell* **85**, 83–94

8. Schweizer, F.E., Dresbach, T., Debello, W.M., O'Connor, V., Augustine, G.J. & Betz, H. (1998) Regulation of neurotransmitter release kinetics by NSF. *Science* **279**, 1203–1206

9. Hanson, P.I., Roth, R., Morisaki, H., Jahn, R. & Heuser, J.E. (1997) Structure and conformational changes in NSF and its membrane receptor complexes visualized by quick-freeze/deep-etch electron microscopy. *Cell* **90**, 523–535

10. Weber, T., Zemelman, B.V., McNew, J.A., Westermann, B., Gmachl, M.J., Parlati, F., Söllner, T.H. & Rothman, J.E. (1998) SNAREpins: minimal machinery for membrane fusion. *Cell* **92**, 759–772

11. Geppert, M. & Südhof, T.C. (1998) Rab3 and synaptotagmin – the yin and yang of synaptic membrane fusion. *Annu. Rev. Neurosci.* **21**, 75–95

12. Bean, A.J. & Scheller, R.H. (1997) Better late than never: a role for rabs late in exocytosis. *Neuron* **19**, 751–754

13. Martin, T.F.J. (1997) Phosphoinositides as spatial regulators of membrane traffic. *Curr. Opin. Neurobiol.* **7**, 331–338

14. Ann, K., Kowalchyk, J.A., Loyet, K.M. & Martin, T.F. (1997) Novel Ca^{2+}-binding protein (CAPS) related to UNC-31 required for Ca^{2+}-activated exocytosis. *J. Biol. Chem.* **272**, 19637–19640

15. Schiavo, G., Gu, Q.-M., Prestwich, G.D., Söllner, T.H. & Rothman, J.E. (1996) Calcium-dependent switching of the specificity of phosphoinositide binding to synaptotagmin. *Proc. Natl. Acad. Sci. U.S.A.* **93**, 13327–13332

16. Baines, A.J., Chan, K.M. & Goold, R. (1995) Synapsin I and the cytoskeleton: calmodulin regulation of interactions. *Biochem. Soc. Trans.* **23**, 65–70

17. Benfenati, F. & Valtorta, F. (1995) Neuroexocytosis. *Curr. Top. Microbiol. Immunol.* **195**, 195–219

18. Rosahl, T.W., Spillane, D., Missler, M., Herz, J., Selig, D.K., Wolff, J.R., Hammer, R.E., Malenka, R.C. & Südhof, T.C. (1995) Essential functions of synapsins I and II in synaptic vesicle regulation. *Nature (London)* **375**, 488–493

19. Prekeris, R. & Terrian, D.M. (1997) Brain myosin V is a synaptic vesicle-associated motor protein: evidence for a Ca^{2+}-dependent interaction with the synaptobrevin–synaptophysin complex. *J. Cell Biol.* **137**, 1589–1601

20. Cremona, O. & De Camilli, P. (1997) Synaptic vesicle endocytosis. *Curr. Opin. Neurobiol.* **7**, 323–330

21. Fesce, R., Grohovaz, F., Valtorta, F. & Meldolesi, J. (1994) Neurotransmitter release: fusion or "kiss-and-run". *Trends Cell. Biol.* **4**, 1–4

22. TerBush, D.R., Maurice, T., Roth, D. & Novick, P. (1996) The exocyst is a multiprotein complex required for exocytosis in *Saccharomyces cerevisiae*. *EMBO J.* **15**, 6483–6494

23. Kee, Y., Yoo, J.S., Hazuka, C.D., Peterson, K.E., Hsu, S.C. & Scheller, R.H. (1997) Subunit structure of the mammalian exocyst complex. *Proc. Natl. Acad. Sci. U.S.A.* **94**, 14438–14443

4

Mitochondria in the life and death of neurons

Samantha L. Budd[*] and David G. Nicholls[†][1]

CNS Research Institute, LMRC First Floor, Brigham & Women's Hospital, Harvard Medical School, 221 Longwood Avenue, Boston, MA 02115, U.S.A., and †Pharmacology and Neuroscience, Ninewells Medical School, University of Dundee, Dundee DD1 9SY, Scotland, U.K.

Introduction

Mitochondria provide the majority of the ATP required for normal neuronal function. In accordance with this energetic role, a chronic decrease in mitochondrial ATP-generating capacity resulting either from genetic alterations to components of the mitochondrial respiratory chain or ATP synthase, or from dysfunction of any of these complexes, is implicated in the pathogenesis of aging and neurodegenerative disorders. Additionally, acute disruption of mitochondrial bioenergetics resulting from Ca^{2+} overload and/or generation of reactive oxygen species plays a major contributory role in the cell death associated with brain ischaemia.

In addition to their principal function of ATP production, it has been discovered that mitochondria participate directly in the induction of programmed cell death (apoptosis) by releasing pro-apoptotic proteins. This pro-active role for mitochondria in a cell death pathway may link the competence of mitochondria to produce ATP with the decision to enter the apoptotic pathway. The present article will attempt to outline the significance of mitochondrial dysfunction in a number of neuronal disease states.

---

[1]*To whom correspondence should be addressed.*

Bioenergetic functions of brain mitochondria

Neurons, in common with most cell types, expend energy to combat the constant inward leak of Na^+ and Ca^{2+}. However, along with other excitable cells, neurons additionally need to restore their ion gradients following controlled fluxes of ions across the cell membrane during electrical signalling. The task of maintaining the ionic homoeostasis in neurons falls largely to ATP-consuming plasma membrane ion pumps such as the Na^+, K^+-ATPase and the Ca^{2+}-ATPase. In fact, most of the ATP consumed by these cells is utilized by these pumps and is generated by mitochondria.

The mitochondrial proton circuit provides the link between the energy-yielding transfer of electrons from NADH or $FADH_2$ to the final electron acceptor (O_2) and the energy-requiring synthesis of ATP (Figure 1). Three of the respiratory chain complexes (I, III and IV) translocate protons from the matrix across the mitochondrial inner membrane into the inter-membrane

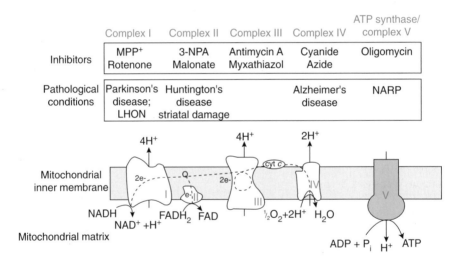

Figure 1. Mitochondrial respiratory chain inhibitors mimic neurodegenerative disorders
The mitochondrial respiratory chain consists of five multisubunit protein complexes (complexes I–V, also known as NADH-UQ oxidoreductase, succinate dehydrogenase, UQ-cytochrome c oxidoreductase, cytochrome c oxidase and ATP synthase respectively) plus two mobile electron carriers, ubiquinone (Q) and cytochrome c (cyt c). The electron transport chain (complexes I–IV, ubiquinone and cytochrome c) carry reducing equivalents from NADH and $FADH_2$ in sequence from ubiquinone to complex III, cytochrome c, complex IV and finally to oxygen, which is reduced to water. The extrusion of protons (H^+) by complexes I, III and IV is coupled to the movement of electrons, generating a proton electrochemical gradient. Protons re-enter the matrix through the ATP synthase to generate ATP. Numerous toxins are known to interfere specifically with mitochondrial respiratory chain function: 1-methyl-4-phenylpyridinium (MPP^+) and 3-nitropropionic acid (3-NPA) have led to primate models of Parkinson's disease and Huntington's disease respectively. Inhibition of complex III is not currently associated with any known neurological condition, although it is inhibited by the pro-apoptotic substance, ceramide. Among the large number of mitochondrial genetic disorders are Leber's Hereditary Optic Neuropathy (LHON), which affects a subunit of complex I, and neuropathy, ataxia and retinitis pigmentosa (NARP), which affects a subunit of the ATP synthase. For a comprehensive review, see [26].

space, generating an electrochemical gradient of protons. This is referred to as the proton motive force (Δp), and consists of two components: ΔpH, the concentration difference of protons across the membrane, and $\Delta\psi$, the membrane potential [1]. The proton circuit is completed by the ATP synthase (also known as complex V), which couples the re-entry of protons through the inner membrane to the synthesis of ATP from ADP and P_i. The $\Delta\psi$ component is also utilized to accumulate Ca^{2+} from the cytoplasm whenever its concentration rises above a 'set-point' of about $0.3-0.5$ μM [1]. Complex III, and to a lesser extent complex I, are major sites of generation of potentially toxic reactive oxygen species [2].

Mitochondrial ATP synthesis is a complex and fragile process which is vulnerable to a large number of factors. In addition to direct inhibition by acute substrate or oxygen deprivation, more chronic effects can be caused by Ca^{2+} overload or cumulative oxidative damage, which may increase the permeability of the inner membrane to protons and/or inhibit mitochondrial metabolism [3]. Current research is focused on two interrelated principles: that cells may die if their capacity for ATP synthesis falls below their peak ATP demand, and that damaged mitochondria may trigger the programmed destruction of the cell.

Mitochondria and excitotoxicity

Many studies show that in acute neurological injuries, such as ischaemia or head trauma, glutamate is released during the insult and the ensuing neuronal death is associated with pathological, excitotoxic activation of N-methyl-D-aspartate (NMDA) receptors [4]. Failure of mitochondrial ATP synthesis in ischaemic neurons leads to a decline in plasma membrane Na^+ and K^+ gradients and a massive efflux of glutamate, mostly of cytoplasmic origin [5]. The combination of high extracellular glutamate and lowering of the neuronal plasma membrane potential, which will alleviate the potential-dependent Mg^{2+} block of the NMDA receptor ion channel, leads to a prolonged activation of the receptor and consequent Ca^{2+} entry across the postsynaptic membrane. This will proceed after re-oxygenation until the plasma membrane is repolarized and glutamate is re-accumulated by the cells. There is overwhelming evidence that this Ca^{2+} entry is responsible for the initiation of the neuron's death [4]. During this process the mitochondria accumulate large amounts of Ca^{2+}. Under certain conditions, isolated mitochondria respond to massive Ca^{2+} loading with the production of reactive oxygen species and a catastrophic permeabilization of their inner membrane, the 'mitochondrial permeability transition' [3]. It has been proposed that this latter event may trigger both programmed (apoptotic) and necrotic neuronal cell death [6]. Thus inhibitors which prevent mitochondria from accumulating Ca^{2+} protect cells *in vitro* against glutamate excitotoxicity [7].

The role of these mechanisms in the much slower, selective neuronal cell death characteristic of neurodegenerative disorders is more speculative. However, in general, relatively mild defects in neuronal metabolism may predispose neurons to NMDA-receptor-mediated excitotoxicity in response to otherwise sublethal concentrations of glutamate [8].

Mitochondrial hypotheses for neurodegenerative disorders

Disorders in oxidative phosphorylation caused by mutations in mitochondrial DNA (mtDNA) or the effects of mitochondrial inhibitors (see Figure 1) suggest that disturbances in mitochondrial oxidative phosphorylation may play a role in a variety of degenerative processes, such as Parkinson's disease (PD), Alzheimer's disease (AD), Huntington's disease (HD) and amyotrophic lateral sclerosis. Indeed, a gradual electron transport 'run-down' may be associated with normal aging. mtDNA accumulates mutations faster than does chromosomal DNA, and this is presumed to be due to oxygen radical-induced damage during the lifetime of an individual. Thus accumulation of mtDNA mutations in post-mitotic senescent tissues may result in bioenergetic deficiency [9]. As discussed above, the mitochondria themselves are an important source of damaging oxygen radicals and are protected from them by various antioxidant systems. Neurodegenerative disorders are characterized by a slow onset and often progressive loss of certain populations of neurons. Neurons are particularly sensitive to inhibition of oxidative phosphorylation, and this can be demonstrated by the ability of mitochondrial toxins to induce neuronal cell death [8]. Impairment of particular sites within the respiratory chain is thought to underlie the differences in the neurodegenerative disorders (Figure 1), although all points of inhibition would be expected to result in a decreased capacity for production of ATP. A sensitization to glutamate excitotoxicity may underlie many of the neurodegenerative consequences of these mitochondrial lesions [8].

Parkinson's disease

PD is a common, late-onset disease associated with movement disorders and loss of dopaminergic neurons in the substantia nigra. Several factors are implicated in the pathogenesis of PD; these include increased deposition of iron in the substantia nigra, enhanced oxidative stress due to monoamine oxidase activity and respiratory chain inhibition, genetic predisposition, and weak excitotoxicity as a consequence of defects in oxidative phosphorylation [10]. While each of these can be demonstrated to play a part in the selective destruction of dopaminergic neurons in PD, and are not necessarily mutually exclusive, the weak excitotoxicity theory, which depends upon a chronic decrease in ATP, currently receives the most attention [10]. Mitochondrial defects have been studied extensively in tissues from PD patients and recently also in cybrid cells [cells in which the native mtDNA has been destroyed, usually by ethidium bromide treatment, and the cells repopulated with the mutant mitochondria from the patient, usually from platelets (since these cells lack nuclear DNA)]. Thus mtDNA from a PD patient inserted into a neuroblastoma cell line resulted in decreased complex I activity [11]. However, other studies have not observed detectable changes in complex I proteins in PD.

Inhibition of complex I with either rotenone or 1-methyl-4-phenylpyri-dinium (MPP^+) [the metabolite of 1-methyl-4-phenyl-1,2,3,6-tetrahydropyri-dine (MPTP)] has been shown to cause the selective loss of dopaminergic neurons when injected *in vivo* [12]. Furthermore, the development of severe PD-like symptoms by intravenous drug users was associated with contamination of the narcotics by MPTP. It was subsequently demonstrated that MPTP is oxidized to MPP^+ by monoamine oxidase B, found in both glial cells and serotonergic neurons in the brain, and then actively transported into nigrostriatal dopaminergic neurons by the dopaminergic uptake system. Once inside these neurons, MPP^+ is then accumulated within the mitochondrial matrix in response to $\Delta\psi$, where it exerts its inhibitory effect on complex I.

In addition to these studies with poisoned mitochondria, deficiencies in oxidative phosphorylation have been found in the platelets, muscle and brain of PD patients. A decrease in complex I activity of 26% was found in the cybrids formed from the combination of PD platelet mtDNA and a neuroblastoma cell line depleted of mtDNA [11]. The PD cybrids were also found to be less effective at restoring the basal cytoplasmic free Ca^{2+} concentration following a Ca^{2+} load. However, bypassing complex I by donating electrons into the respiratory chain at complex II using succinate (Figure 1) could partially restore PD cybrid Ca^{2+} recovery [11]. The fact that PD patient platelet and muscle mitochondria demonstrate a deficit in complex I activity would seem to indicate the presence of a mtDNA lesion rather than the specific accumulation of an environmental toxin by the affected neurons. However, this is an oversimplification, and the symptoms of PD appear to be biochemically and genetically heterogeneous, indicating both genetic and environmental inhibition of oxidative phosphorylation.

The inhibition of complex I *per se* appears to be accompanied by the generation of reactive oxygen species such as superoxide [13], which may additionally cause oxidative damage to the other respiratory chain complexes. Interestingly, inhibition of complex III (by antimycin A, but not by myxothiazol) greatly enhances superoxide generation [13], although inhibition of complex III is currently not associated with any known neurological condition. This implies that oxidative damage in PD may be secondary to the respiratory chain deficiency.

Alzheimer's disease

AD is the most common neurodegenerative disease of later life and manifests clinically as a progressive cognitive decline. The primary mechanisms initiating neuronal degeneration in AD are not yet resolved, although several potential pathogenic mechanisms are known. While the primary focus of research has been on the involvement of tau hyperphosphorylation, amyloid β peptide and the amyloid precursor protein [14], there is evidence implicating impaired metabolic function and oxidative stress in AD, with a focus on deficits in complex IV activity (Figure 1). Thus complex IV activity *post mortem* has been reported to be reduced by 25−30% in the frontal, temporal, parietal and occipital cortex of AD brains [15]. In addition, AD cybrids formed from patient platelet mtDNA and a mtDNA-depleted neuroblastoma cell line demonstrated a 52% decrease in complex IV activity compared with control cybrids [16]. However, not all studies have shown mitochondrial defects in AD brain biopsies, and a recent claim that mutations in two of the complex IV genes encoded by mtDNA (CO1 and CO2) were demonstrated to be inherited in a maternal fashion, and to segregate at a higher frequency with AD [17], has been reported to be artifactual [18]. The significance of a complex IV deficiency as a causative factor in AD is thus still controversial.

Huntington's disease

HD is associated with the selective loss of neurons in the basal ganglia. The nuclear gene defect responsible for HD results in an abnormal polyglutamine extension of a protein, huntingtin, of unknown function. While huntingtin is not located in mitochondria, several lines of evidence implicate the involvement of an energetic defect and consequent oxidative damage in HD. Inhibition of complex II by systemic administration of 3-nitropropionic acid in rats and primates (Figure 1) is used to produce animal models displaying lesions that closely resemble the neuropathological features of HD [19]. The lesion induced by systemic or intrastriatal injection of 3-nitropropionic acid is associated with metabolic impairment and, interestingly, is also age-dependent, in that young animals are more resistant to its neurotoxic effects [19].

Mitochondrial encephalopathies, neuropathies and myopathies

The mitochondrion has between two and ten molecules of mtDNA, which encodes 13 polypeptides and all the machinery required for their transcription and translation. Importantly, mtDNA lacks a protective histone coat and the normal repair enzymes associated with nuclear DNA. Thus mtDNA acquires mutations at a much higher rate than nuclear DNA [9]. The 13 polypeptides encoded by mtDNA are all subunits of the respiratory chain complexes; they include six subunits of complex I, cytochrome b, the three larger subunits of complex IV, and one subunit of the ATP synthase. The remaining mitochondrial proteins (90−95% of the total) are encoded by nuclear DNA and are imported post-translationally via specific translocation systems. Clearly, defects in mitochondria- or nuclear-encoded enzymes, or in their transport or assembly, could lead to dysfunction of the respiratory chain. Tissues with high rates of aerobic metabolism, such as brain, retina, and cardiac and skeletal muscle, are particularly vulnerable to decreases in oxidative phosphorylation.

The disorders presented by mutations in mtDNA are particularly interesting from a genetic point of view, firstly because mtDNA is inherited in a maternal fashion, and secondly because there are thousands of copies of mtDNA in each cell, so that the phenotypic expression of a mitochondrial gene will depend on the relative proportions of mutant and wild-type mtDNAs within a cell. The latter, in conjunction with mitochondria replicating in post-mitotic cells, results in heteroplasmy of any mitochondrial mutation within cells.

To date, a number of deleterious mtDNA mutations have been identified, including both maternally inherited point mutations and sporadic large-scale rearrangements. Patients may present at any age with symptoms ranging from acute episodes of lactic acidosis in infancy to severe neurodegenerative illness in adulthood. Unfortunately, while advances in molecular genetics techniques have facilitated the diagnosis and characterization of specific molecular defects, there is a lack of definitive biochemical therapies. In most cases, the phenotypic expression of the mtDNA mutation varies depending on the degree of mitochondrial heteroplasmy within different tissues of the same individual (segregative replication).

Mitochondria and programmed cell death

A chronic decrease in ATP and overstimulation of excitatory receptors may be associated with the above neurodegenerative disorders, but how the neurons die following this is not entirely clear. It is known that apoptosis (programmed cell death) is an ATP-requiring pathway, and that if cellular ATP levels fall too low during a death-inducing stimulation then necrosis (disordered cell death) ensues [20]. Mitochondrial dysfunction is actively involved in the initiation of apoptosis, and mitochondria additionally participate in the signalling pathway that culminates in nuclear chromatin condensation. An initiating signal for apoptosis appears to be incipient damage in a critical proportion of the cell's mitochondria, signalled by their release of pro-apoptotic factors, rather than a subsequent collapse in ATP. Indeed, if neurons or human T-cells [20] are depleted of ATP, apoptosis is no longer possible, and necrosis ensues.

The protein Bcl-2 is a potent inhibitor of apoptosis induced by various signals, and is mainly localized to the mitochondrial outer membrane. The mechanism by which Bcl-2 functions is still controversial. Antioxidant or pore formation have been proposed, although the clearest phenotype observed with mitochondria isolated from a neural cell line overexpressing Bcl-2 is a decreased susceptibility to the permeability transition [21].

Mitochondrial swelling as a consequence of the permeability transition ruptures the outer membrane and allows release of pro-apoptotic factors from the intermembrane space. To date, two such factors have been proposed: the apoptosis-inducing factor (AIF) [22] and cytochrome c [23]. These factors appear to differ in the way that they effect the end stages of apoptosis (see Figure 2), but the release of both AIF and cytochrome c from the intermembrane space appears to be under the control of Bcl-2 [22,23]. On release from the mitochondrion, cytochrome c, which normally functions as a mobile transporter of electrons between complexes III and IV (see Figure 1), participates in the activation of caspases implicated in apoptosis [23]. In a cell-free system purified cytochrome c, together with dATP, triggers the activation of Apaf-1 (apoptosis-activating factor-1). This then cleaves and activates caspase-9, which in turn activates caspase-3 (CPP32) [24]. These caspases are responsible for the

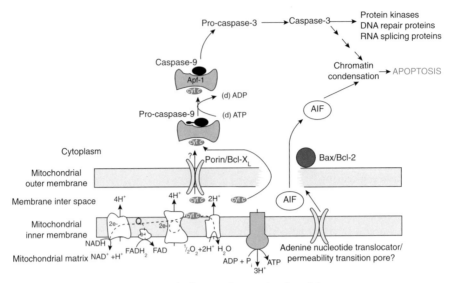

Figure 2. Release of pro-apoptotic factors from mitochondria
Release of cytochrome c (cyt c) from the mitochondria is required for the activation of down-stream caspases. It is unclear whether release is specific via a porin or occurs as a result of outer membrane breakage following matrix swelling initiated by activation of the permeability transition pore (in turn a consequence of oxidative damage and Ca^{2+} overload). Cleavage and activation of caspase-3 is associated with the end-point of apoptosis (chromatin condensation). Caspase-3 cleaves a number of targets and activates a specific DNase. The precise identity of the mitochondrial permeability transition pore remains elusive, but it may be composed of a dimerized adenine nucleotide translocator on the inner membrane, plus porin and an 18 kDa protein in the outer membrane.

key proteolytic events characterizing apoptosis. AIF appears to act further downstream, is itself capable of causing nuclear chromatin condensation, and is insensitive to inhibitors of caspase-3 [22].

Considerable uncertainty surrounds the mechanism for the release of AIF and cytochrome c from the intermembrane space into the cytosol (see Figure 2). While the release of both is under the control of the Bcl-2 protein(s), the release of AIF has been strongly correlated with the opening of the mitochondrial permeability transition pore, depolarization of the mitochondrial membrane, an increase in mitochondrial volume and rupture of the mitochondrial outer membrane [22]. Kroemer and colleagues [22] propose that AIF can then leave the intermembrane space in a non-specific fashion. In contrast, the release of cytochrome c has been reported to occur prior to a decrease in mitochondrial membrane potential, and to be unaffected by inhibitors of pore opening [25].

Perspectives

The mitochondrion occupies the centre stage in current theories of both acute and chronic neurodegeneration. With respect to the former, the development of drugs to control the uptake of Ca^{2+} by mitochondria during post-ischaemic reperfusion, or to limit the damage to mitochondrial bioenergetics resulting from such accumulation, may provide an alternative to current therapies. With respect to chronic neurodegenerative disease, the advent of cybrid technology allows functional investigations of human mitochondrial mutations to be carried out, and this may allow the development of patient-specific therapy regimes. For instance, if a metabolic defect is found, then perhaps agents that can improve oxidative phosphorylation downstream of this defect may be useful. In addition, identifying mtDNA mutations will open up the possibility of gene therapy for neurodegenerative disorders. Certainly, at the present time, if mutations in either mtDNA or nuclear DNA which predispose individuals to degenerative diseases can be found, they will enable the development of presymptomatic disease testing and genetic counselling. Interfering with an apoptotic cascade may only serve to rescue irreparably damaged neurons, and thus be of little value in the therapy of neurodegenerative disorders. However, if upstream mitochondrial dysfunction is responsible for the cell damage that initiates subsequent apoptosis to remove the defective cell, then it is apparent that an ability to prevent this dysfunction might prevent the subsequent cell destruction. Conversely, facilitation of mitochondrially induced apotosis in cancer may contribute to elucidating the close interrelationship between defective cell-cycle control and failed apoptosis in these conditions.

Summary

- *A prolonged decrease in ATP levels underlies a number of neurodegenerative disorders.*
- *Defects in oxidative phosphorylation are associated with a number of neurodegenerative disorders.*
- *Mitochondria also play an important role in mediating the initiation of apoptosis.*

References

1. Nicholls, D.G. & Ferguson, S.J. (1992) Bioenergetics 2, Academic Press, London
2. Dykens, J.A. (1994) Isolated cerebral and cerebellar mitochondria produce free radicals when exposed to elevated Ca^{2+} and Na^+: implications for neurodegeneration. J. Neurochem. 63, 584–591
3. Bernardi, P. & Petronilli, V. (1996) The permeability transition pore as a mitochondrial calcium release channel: a critical appraisal. J. Bioenerg. Biomembr. 28, 131–138
4. Choi, D.W. (1996) Ischemia-induced neuronal apoptosis. Curr. Opin. Neurobiol. 6, 667–672
5. Nicholls, D.G. & Attwell, D.A. (1990) The release and uptake of excitatory amino acids. Trends Pharmacol. Sci. 11, 462–468
6. Petit, P.X., Susin, S.A., Zamzami, N., Mignotte, B. & Kroemer, G. (1996) Mitochondria and programmed cell death: back to the future. FEBS Lett. 396, 7–13
7. Budd, S.L. & Nicholls, D.G. (1996) Mitochondrial calcium regulation and acute glutamate excitotoxicity in cultured cerebellar granule cells. J. Neurochem. 67, 2282–2291
8. Henneberry, R.C. (1997) Excitotoxicity as a consequence of impairment of energy metabolism: the energy-linked excitotoxic hypothesis. In Mitochondria and Free Radicals in Neurodegenerative Disease (Beal, M.F., Howell, N. & Bodis-Wollner, I., eds.), pp. 111–143, Wiley–Liss, New York

9. Ozawa, T. (1997) Genetic and functional changes in mitochondria associated with aging. *Physiol. Rev.* **77**, 425–464

10. Blandini, F., Porter, R.H.P. & Greenamyre, J.T. (1996). Glutamate and Parkinson's disease. *Mol. Neurobiol.* **12**, 73–94

11. Sheehan, J.P., Swerdlow, R.H., Parker, W.D., Miller, S.W., Davis, R.E. & Tuttle, J.B. (1997) Altered calcium homeostasis in cells transformed by mitochondria from individuals with Parkinson's disease. *J. Neurochem.* **68**, 1221–1233

12. Cooper, J.M. & Schapira, A.H.V. (1997). Mitochondrial dysfunction in neurodegeneration. *J. Bioenerg. Biomembr.* **29**, 175–183

13. Richter, C., Gogvadze, V., Laffranchi, R., Schlapbach, R., Schweizer, M., Suter, M., Walter, P. & Yaffee, M. (1995) Oxidants in mitochondria: from physiology to diseases. *Biochim. Biophys. Acta* **1271**, 67–74

14. Yankner, B.A. (1996) Mechanisms of neuronal degeneration in Alzheimer's disease. *Neuron* **16**, 921–932

15. Sheehan, J.P., Swerdlow, R.H., Miller, S.W., Davis, R.E., Parks, J.K., Parker, W.D. & Tuttle, J.B. (1997) Calcium homeostasis and reactive oxygen species production in cells transformed by mitochondria from individuals with sporadic Alzheimer's disease. *J. Neurosci.* **17**, 4612–4622

16. Mutisya, E.M., Bowling, A.C. & Beal, M.F. (1994) Cortical cytochrome oxidase activity is reduced in Alzheimer's disease. *J. Neurochem.* **63**, 2179–2184

17. Davis, R.E., Miller, S., Herrnstadt, C., Ghosh, S.S., Fahy, E., Shinobu, L.A., Galasko, D., Thal, L.J., Beal, M.F., Howell, N. & Parker, Jr., W.D. (1997) Mutations in mitochondrial cytochrome c oxidase genes segregate with late-onset Alzheimer disease. *Proc. Natl. Acad. Sci. U.S.A.* **94**, 4526–4531

18. Hirano, M., Shtilbans, A., Mayeux, R., Davidson, M.M., DiMauro, S., Knowles, J.A. & Schon, E.A. (1997) Apparent mtDNA heteroplasmy in Alzheimer's disease patients and in normals due to PCR amplification of nucleus-embedded mtDNA pseudogenes *Proc. Natl. Acad. Sci. U.S.A.* **94**, 14894–14899

19. Beal, M.F., Brouillet, E., Jenkins, B.G., Ferrante, R.J., Kowall, N.W., Miller, J.M., Storey, E., Srivastava, R., Rosen, B.R. & Hyman, B.T. (1993) Neurochemical and histological characterization of striatal excitotoxic lesions produced by the mitochondrial toxin 3-nitropropionic acid. *J. Neurosci.* **13**, 4181–4192

20. Leist, M., Single, B., Castoldi, A.F., Khunle, S. & Nicotera, P. (1997) Intracellular adenosine triphosphate (ATP) concentration: a switch in the decision between apoptosis and necrosis. *J. Exp. Med.* **185**, 1481–1486

21. Murphy, A.N., Bredesen, D.E., Cortopassi, G., Wang, E. & Fiskum, G. (1996) Bcl-2 potentiates the maximal calcium uptake capacity of neural cell mitochondria. *Proc. Natl. Acad. Sci. U.S.A.* **93**, 9893–9898

22. Susin, S.A., Zamzami, N., Castedo, M., Hirsch, T., Marchetti, P., Macho, A., Daugas, E., Geuskens, M. & Kroemer, G. (1996) Bcl-2 inhibits the mitochondrial release of an apoptotic protease. *J. Exp. Med.* **184**, 1331–1341

23. Liu, X., Kim, C.N., Yang, J., Jemmerson, R. & Wang, X. (1996) Induction of apoptotic program in cell-free extracts: requirement for dATP and cytochrome c. *Cell* **86**, 147–157

24. Li, P., Nijhawan, D., Budihardo, I., Srinivasula, S.M., Ahmad, M., Alnemri, E.S. & Wang, X. (1997) Cytochrome c and dATP-dependent formation of Apaf-1/caspase-9 complex initiates an apoptotic protease cascade. *Cell* **91**, 479–489

25. Yang, J., Liu, X., Bhalla, K., Kim, C.N., Ibrado, A.M., Cai, J., Peng, T.-I., Jones, D.P. & Wang, X. (1997) Prevention of apoptosis by Bcl-2: release of cytochrome c from mitochondria blocked. *Science* **275**, 1128–1132

26. Beal, M.F., Howell, N. & Bodis-Wollner, I. (eds.) (1997) *Mitochondria and Free Radicals in Neurodegenerative Diseases*, Wiley–Liss, New York

5

Neuro-regeneration: plasticity for repair and adaptation

Pico Caroni

Friedrich Miescher Institute, P.O. Box 2543, CH-4002 Basel, Switzerland

Introduction

In order to process information in a manner that is meaningful and useful to the organism, the nervous system is 'wired' with remarkable precision. It now seems that the precision of neuronal connections is due largely to the early acquisition of distinct phenotypes by differentiating neurons, which in turn are predictive of neuronal projections and connectivity [1–3]. These neuronal phenotypes are thought to be determined by defined combinations of transcription factors. In olfactory neurons, for example, predetermination extends to the level of groups of neurons expressing single specific receptors, whereas, for spinal cord motoneurons, identity may define groups of neurons with specific neurotransmitter and electrophysiological properties that project to the same muscle, or even to a particular subcompartment of the same muscle. Just as neurons appear to know early on who they are and how they are to connect, they also express sets of genes associated with axon, dendrite and synapse formation at defined times during development. In particular, genes associated with axon formation and elongation are down-regulated either when the target region is reached or when the process of target innervation is completed [4]. The significance of these identity- and intrinsic programme-based properties of the neuronal phenotype for neuro-regeneration is that appropriate programmes of gene expression may have to be re-activated in adult neurons in order to allow regeneration [5,6]. In addition, regeneration and plasticity are likely to be influenced by differences in neuronal phenotype.

Although the major phases of wiring and synapse formation are restricted to specific stages of neural development, the adult nervous system is capable of dramatic activity-regulated plasticity. Ongoing physiological plasticity in the adult is essential for learning and memory. The corresponding long-term changes in circuit activity and synaptic strength probably involve persistent alterations in the molecular composition of affected synapses, but may also involve alterations and remodelling of synaptic structures. In addition, the adult nervous system can carry out extensive re-organization of synaptic circuits in order to adapt to drastic alterations in the pattern of input activity. One well-characterized example is the induction of small experimental scotomas in the eye [7,8]. In these experiments, a circumscribed patch of photoreceptor cells is lesioned, without damaging retina neurons and in particular the retinal ganglion cells that carry visual information from the eye to the brain. Due to the particular synaptic organization of the visual system, such lesions result in the electrical silencing of a corresponding patch of retinal ganglion cells in the neuroretina and of cortical neurons in the primary visual cortex. As a consequence of this inactivation, adjacent projections in the visual cortex expand, the affected cortical patch becomes driven by these neighbouring inputs and the map of the visual world in the primary visual cortex acquires a corresponding distortion. In the visual cortex the blind spot in the visual field is refilled and, depending on the size of the scotoma, the animal can learn to reconstruct and process accurate images of the visual world, i.e. it adapts. Such rearrangements can be reversible; they can also have dramatic non-adaptive consequences, as seen, for example, in phantom limb pain. The mechanisms underlying this plasticity involve the activation of pre-existing but largely silent synaptic connections [7], and possibly also local nerve sprouting and the formation of new synaptic connections [8]. In the context of neuro-regeneration, these findings emphasize the fact that neuronal circuits have a strong capacity for functional rearrangements in the adult. Important issues include the mechanisms that allow, promote and regulate the specificity of such plasticity. Furthermore, are there plastic states in neurons just like there are axon elongation modes, and are there differences in the plasticity competence of neuron types and neural networks?

A critical aspect of neuro-regeneration is the long-distance regrowth of lesioned axons in order to restore functional connections between processing areas that have been disconnected by the lesion. This regenerative process is efficient during development, when gene expression for axonal growth is turned on and the local environment is favourable. In lower vertebrates, efficient regeneration of lesioned axons is also observed in the adult. In adult mammals, however, such regeneration is essentially restricted to the peripheral nervous system (PNS) (Figure 1). Since there may have been little evolutionary pressure to maintain axonal regeneration in the adult central nervous system (CNS), suppression of axonal regeneration may have evolved as a mechanism to confine structural plasticity to appropriate target regions. In addition to axonal regener-

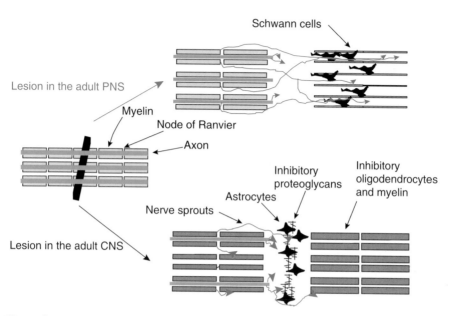

Figure 1. Factors in the local environment that promote the regeneration of lesioned axons in the adult PNS, and inhibit it in the adult CNS

In the PNS, myelin in the distal nerve is rapidly degraded, and Schwann cells are activated to express factors that promote axon growth. Collateral sprouts from nodes of Ranvier rapidly cross the lesion site and elongate inside empty basal lamina sheets. In the CNS, myelin degradation is slow, several neurons die as a consequence of the lesion, and scar-associated astrocytes deposit inhibitory proteoglycans into the extracellular matrix; in addition, inhibitory components on the surface of oligodendrocytes and myelin prevent regeneration.

ation, reactive nerve sprouting contributes to local neuro-regeneration both in the PNS and in the CNS. Finally, to have a functional impact, neuro-regeneration must involve the plasticity of pre-existing synapses and the formation of new synaptic connections. This brief overview will focus on some of the mechanisms that control the capacity of the nervous system for repair.

Reactions of adult neurons to axotomy

A lesion of the axon (axotomy) is a traumatic event for neurons that can initiate major changes in gene expression in the cell body [5,6]. Except for a brief period late in development, when for poorly understood reasons lesions induce massive neuronal death, PNS neurons react to axotomy by a switch in their biosynthetic machinery and the re-expression of genes associated with axon formation (see below). This biosynthetic switch or chromatolysis reaction involves a major rearrangement of the endoplasmic reticulum and Golgi apparatus that is presumably related to the transition from a secreting to a growing cell. The signals that initiate this switch in neuronal phenotype are not understood, but may involve a lesion-induced interruption of retrograde inhibitory signalling from the axon periphery to the cell body. In marked contrast with PNS neurons, axotomy in adult CNS neurons of higher vertebrates can induce substantial cell death [9]. Interestingly, susceptibility to axotomy-induced cell death varies significantly among different CNS neurons, is much more pronounced when axons are lesioned in the vicinity of the cell body, and seems to correlate with the intensity with which the cell body responds to the injury [10]. Unfortunately, reaction intensity also correlates with the potential of CNS neurons to initiate axonal regeneration [10]. As discussed below, this paradoxical relationship between susceptibility to apoptotic cell death and the transition to an axonal growth mode is a major obstacle to axonal regeneration in the adult CNS in higher vertebrates.

Axonal growth during development and regeneration correlates with the expression of a distinct group of genes in neurons. The term 'growth-associated proteins' (GAPs) initially referred to proteins that are greatly induced and rapidly transported in regenerating nerves, thus possibly playing a direct role in axon elongation [5]. The neural protein GAP-43 [4] is possibly the best known example of such a GAP. GAP-43 is a major protein kinase C substrate and calmodulin-binding protein in the brain. It accumulates selectively at the plasma membrane of axons, where it affects growth cone activity and vesicle fusion through mechanisms that have not yet been elucidated. The current list of known GAPs includes a variety of vesicle- and/or membrane-associated proteins, including cytoskeleton-regulating proteins (e.g. SCG-10), cell adhesion molecules [e.g. N-CAM (neural cell adhesion molecule)], receptors for extracellular-matrix proteins (e.g. β_1-integrin), the low-affinity nerve growth factor receptor, and non-receptor tyrosine kinases (e.g. c-Src). Further genes and proteins associated with axon growth include cytoskeletal proteins such as actin and tubulin isoforms, and transcription factors such as c-Jun [5,10].

A combination of PNS-graft regeneration experiments (see also next section) and the application of molecular markers for axonal growth has provided valuable insights into the possible relationships between the expression of GAP genes and the actual growth process. These experiments have revealed a strong correlation between the expression of GAP-43 and competence for axonal growth. In one set of experiments, lesioned optic nerve axons only grew into peripheral nerve grafts when a lesion was applied within about 1 mm of the eye [11]. Unfortunately, such lesions induce massive death of retinal ganglion cells, i.e. the neurons that convey all information from the eye to the brain. Lesions further into the CNS had far less devastating consequences for retinal ganglion cell survival, but resulted in the complete absence of axonal regeneration into grafts [12]. Apparently, an intense cell body reaction is required for regeneration, but this can also lead to apoptosis. GAP-43 expression was only detected in a subpopulation of lesioned ganglion cells, and only with proximal lesions [12]. Labelling of regenerating axons with lipophilic dyes that diffuse back into the cell body (retrograde labelling) revealed that the few ganglion cells that extended axons into the grafts all expressed elevated levels of GAP-43, suggesting that a cell body reaction visualized by the expression of this GAP may be causally associated with successful axon elongation. However, with such lesions GAP-43 induction was also detected in the absence of the graft [13], and thus of effective regeneration. Many neurons that expressed GAP-43 did not regenerate, indicating that the expression of GAP-43, and presumably also of further growth-related genes, is not sufficient for nerve regeneration.

Strong evidence in support of a link between a cell body reaction (defined by the induction of GAP-43) and the vigour of axon elongation into a peripheral nerve graft was provided by experiments with dorsal root ganglion (DRG) neurons. The axon of these PNS neurons bifurcates close to the cell body, giving rise to branches growing to the periphery and into the CNS. Peripheral root lesions lead to robust induction of GAP-43 and nerve regeneration, whereas dorsal root lesions lead to very poor regeneration, no growth of axons into the spinal cord and no detectable induction of GAP-43. Application of a peripheral nerve graft next to a lesioned dorsal root results in the slow growth of processes into the graft, and in a delayed but significant induction of GAP-43 [14,15]. Crushing of the peripheral nerve branch of the same DRG neurons at the time of nerve grafting to their dorsal root led to greatly accelerated process growth into the graft and to rapid induction of GAP-43. Therefore a vigorous cell body reaction induced by a peripheral root lesion and reflected by the induction of GAP-43 mRNA significantly accelerates the elongation of dorsal root axons in the nerve graft [15]. While the peripheral nerve lesion induced rapid and transient GAP-43 expression in a very large number of DRG neurons, it did not lead to a comparable elevation in the numbers of dorsal root axons growing into the peripheral nerve graft [15]. As with the retinal ganglion cell experiments mentioned above, the peripheral nerve lesion apparently affected growth vigour, but not the absolute numbers of dorsal root axons able to regenerate into the graft. It will be important to determine whether the absence of regeneration in lesioned and GAP-re-expressing CNS neurons is due to some particular intrinsic properties of these neurons or, alternatively, to lesion-related unfavourable properties of their local environment.

Role of extrinsic factors in axonal regeneration

Pioneering experiments by Albert Aguayo's group have established that the local environment in the adult CNS of higher vertebrates plays a major role in preventing axonal regeneration after a lesion [16] (Figure 2). Thus not only do central neurons regenerate axons into a peripheral nerve graft, but peripheral axons fail to regenerate axons into a central nerve graft. Several factors are responsible for these differences. The main growth-promoting activity in peripheral nerve explants is due to living Schwann cells activated by the absence of axon contact. In contrast, CNS nerves prevent axon growth due to inhibitory activities associated with oligodendrocytes and their product, CNS myelin [17]. In addition, lesioned peripheral nerves contain higher levels of growth-promoting neurotropic factors than do lesioned central nerves. The experiments

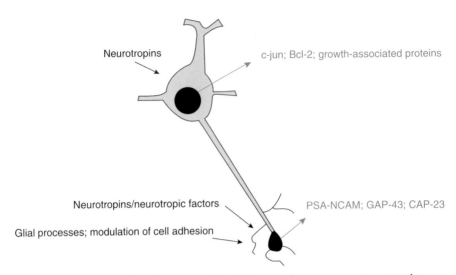

Figure 2. Intrinsic (blue) and extrinsic (black) factors that promote regeneration (top) and local nerve sprouting (bottom) in the adult
Abbreviations: PSA, polysialic acid; CAP-23, cortical cytoskeleton-associated protein of 23 kDa.

with peripheral nerve grafts also revealed the existence of intrinsic differences in the regenerating ability and vigour of lesioned axons. Thus certain types of neurons, e.g. cerebellar Purkinje cells [18], consistently failed to regenerate axons into peripheral nerve grafts.

In addition to differences in growth-promoting and -inhibitory activities in peripheral compared with central nerves, the formation of a glial scar, mainly associated with reactive astrocytes, is a major factor that prevents regeneration in the CNS [19]. Recent studies have provided evidence that it is not the astrocytes themselves, but the appearance of inhibitory proteoglycans in the extracellular matrix associated with scar material, that prevents axon growth. Thus in two models of axonal growth in the adult CNS, i.e. growth upon microlesions and extension of axons by adult DRG neurons implanted into the corpus callosum of an adult rat, failures coincided spatially with the presence of proteoglycans [19]. Remarkably, implanted adult DRG neurons extended long axons into the host CNS white matter, suggesting that inhibitory factors from oligodendrocytes may inhibit but do not suppress axonal elongation (see next section). A recent study suggests that deposition of inhibitory proteoglycans may be triggered by blood-derived factors that would invade the CNS as a consequence of a local breakdown of the blood–brain barrier. Clearly, the identification of signalling pathways that regulate the production of such inhibitory molecules could have major clinical implications for promoting regeneration in the CNS. In addition to the neutralization of oligodendrocyte inhibitors and the prevention of inhibitory proteoglycan production, grafting of glial cells that promote and guide axon growth also promises to provide substantial benefits for regenerating axons. In particular, a recent study provided evidence that glial cells derived from the olfactory system not only greatly promote axon elongation but also help regenerating axons to manage the difficult transition into the CNS environment [20].

Role of intrinsic neuronal components in axonal regeneration

Experimental evidence supporting the view that neurons express axonal growth programmes during development, which are greatly attenuated or absent in the adult, derives from co-culture and transplant experiments that used combinations of neurons and tissues from different developmental stages. In one dramatic type of experiment, dissociated embryonic neurons were transplanted into lesioned adult rat brains [21]. When very young neurons were transplanted, they grew long processes in the pre-lesioned adult host brain. Clearly, this is in marked contrast with the lesioned host neurons. In these experiments, extensive growth was detected along myelinated tracts in the CNS. These experiments also provided evidence that grafts of dissociated embryonic neurons may integrate functionally into the host brain, thus providing a potential strategy to counteract functional deficits in certain neurodegenerative diseases, e.g. Parkinson's disease. The application range of such graft strategies is, however, presently limited by the poor survival of transplanted neurons. The actual intrinsic neuronal mechanisms that allow young neurons to grow in an adult CNS environment are not clear. Neurons may differ in general axonal growth programmes that define an elongation mode or elongation vigour. An alternative view is that neurons differ in their expression profiles for the machinery that detects and amplifies guidance signals from the environment. *In vitro* experiments with dissociated neurons and CNS white matter sections, for example, suggest that young neurons may be less sensitive to inhibitory proteins on the surface of oligodendrocytes [17].

Graft regeneration experiments have provided molecular evidence for at least two classes of genes associated with axonal elongation [13]. Thus, while GAP-43 expression was correlated with elongation competence, levels of β-tubulin transcripts were only elevated when actual elongation into the peripheral nerve graft took place. This is reminiscent of the regulation of these transcripts during development. In this case, tubulin and actin transcript levels correlate with axon elongation, whereas GAP-43 expression can persist at high levels during a protracted period of target innervation, when axons arborize extensively within the target region, where they form increasing numbers of synaptic connections. These findings provide valuable molecular correlates for at least two intrinsic growth states in developing and regenerating neurons.

Explant co-culture experiments using hamster retina (as the source of axons) and midbrain tectum (the target region) have provided an *in vitro* experimental system with which to dissect the contributions of intrinsic neuronal properties and the local environment in axonal regeneration [22]. The main finding of these studies was that, while embryonic (day 15) and day 0 postnatal retinal axons regenerated into tectum of any age, including adult, most axons from day 2 postnatal and older retinae failed to grow into tectal explants. Significantly, even embryonic tectal explants failed to attract axon growth from day 2 postnatal and older retinae. These findings provide evidence for a dramatic developmental decline in the intrinsic ability of retinal axons to elongate in tectal explants. Neurons from older retinae did, however, grow processes inside the retina explants, and it was mainly growth into the target region (i.e. tectal) explants that was absent. Therefore a remaining question is the extent to which these findings reflect differences in axon growth *per se*, as opposed to growth into CNS explants. The molecular mechanisms that control an intrinsic elongation programme in neurons may be linked to the anti-apoptotic protein Bcl-2 [23]. Thus the expression of Bcl-2 was apparently necessary and sufficient to promote the elongation of retinal ganglion cell axons in mice. This effect of Bcl-2 was apparently not simply due to its anti-apoptotic properties, suggesting that a molecular switch to an axonal elongation mode in neurons requires Bcl-2. Overexpression of Bcl-2 failed, however, to completely rejuvenate neurons, as late embryonic and early postnatal neurons failed to grow into adult tectum even in the presence of excess Bcl-2. Elucidation of the relative contributions of further intrinsic growth-regulating mechanisms and inhibitory signals in the adult environment may provide crucial leads towards promoting nerve regeneration in the adult CNS.

Factors that control nerve sprouting and synaptogenesis in the adult

Nerve sprouting is involved in target innervation during development, de-afferentation-induced repair, and possibly also use-related plasticity in the adult [6]. During development, when incoming axons reach their target region they branch extensively to establish a large number of synapses. This process of terminal arborization can be preceded by a waiting period, when no axonal growth is detected. At least in some systems, synaptogenesis proceeds for a protracted period of time, as more target sites are added through the generation of additional postsynaptic neurons and the growth of dendrites. During this process, innervation involves sprouting of collaterals, at a time when axonal elongation and the expression of several GAP genes have subsided. Towards the end of this period excess collaterals are eliminated through an activity-dependent process that produces a quantitative refinement of synaptic connections. Synapse elimination can proceed as new target sites are innervated by sprouting collaterals, i.e. there is some overlap in time between the two processes. This suggests that nerve sprouting and synapse retention are not governed by the same factors.

In the adult PNS and CNS, reactive nerve sprouting is a major adaptational mechanism to compensate for lesion-induced de-afferentation. Possibly due to specific contact-mediated inhibition, and like collateral sprouting during development, sprouting in the adult is confined to specific target areas [24]. In neurodegenerative diseases, re-innervation through sprouts from neighbouring neurons can effectively delay the appearance of detectable deficits. It is presently not clear whether, and to what extent, local nerve sprouting in the adult also contributes to activity-dependent network plasticity in the absence of local de-afferentation. Studies initiated in *Aplysia* and recently extended to rodents have provided evidence that activity-regulated alterations in synaptic structure may be a substrate for persistent modifications of synaptic function [25]. Whether presynaptic expansion at synaptic structures may be related to nerve sprouting, and whether new synaptic connections are formed in the adult in the absence of physical de-afferentation, remain to be determined.

All forms of neurite outgrowth, including terminal arborization and nerve sprouting, are induced and guided by extrinsic signals from the local environment [26]. For example, nerve sprouting *in vivo* can be induced by the local application of neurotropins or other neurotropic factors. Although the actual functional roles of such mechanisms in nerve sprouting and terminal arborization are not yet clear, diffusible factors could trigger sprouting, or even attract sprouts by chemotropic mechanisms. Such a mechanism is likely to induce the sprouting of collaterals that is involved in the innervation of several types of targets during development [26]. In addition to diffusible activities, contact-mediated mechanisms play a critical role in nerve sprouting. This is particularly well documented for the adult neuromuscular junction, where sprouts are induced and guided by the processes of activated Schwann cells [27]. The neuromuscular junction must be viewed as a functional unit involving three cellular elements: the presynaptic motor nerve terminal, the postsynaptic skeletal muscle fibre, and a cluster of synapse-associated glia, the terminal Schwann cells. The involvement of the terminal Schwann cell is based on the facts that it senses and reacts to transmitter release and that it interacts selectively with the synaptic extracellular matrix and nerve processes. Similar cellular arrangements are found at certain types of central synapses, but it is not clear whether astrocytic processes can regulate nerve sprouting in the CNS.

In addition to factors in the environment, intrinsic neuronal components also affect the competence of nerves to sprout in the adult. Thus transgenic mice overexpressing GAP-43 selectively in adult neurons exhibit dramatic spontaneous nerve sprouting [28]. To determine whether GAP-43 produced a true gain-of-function phenotype, toxin- and lesion-induced nerve sprouting was compared in control and transgenic mice. To allow for an assessment of the role of GAP-43 in induced sprouting, the experiments were carried out under conditions that also lead to nerve sprouting in the absence of GAP-43. These experiments revealed a dramatic potentiation of induced sprouting in the presence of GAP-43, indicating that this GAP is an intrinsic determinant of nerve sprouting [28]. Whether neurons that maintain substantial levels of GAP-43 expression in the adult also have a greater potential for nerve sprouting remains to be determined. A similar study provided evidence that CAP-23 (cortical cytoskeleton-associated protein of 23 kDa) is a growth-associated protein with sprout-promoting properties similar, but not identical, to those of GAP-43 [29]. GAP-43 and CAP-23 are locally abundant and highly regulated proteins that share a number of biochemical and cell biological properties, and are expressed in partially overlapping subsets of neurons in the adult. One further neuronal protein that may be viewed as an intrinsic determinant of nerve sprouting is the polysialylated form of the cell adhesion protein N-CAM. The differential expression of such intrinsic determinants of nerve sprouting may affect both the pattern and the regulation of nerve sprouting during development and in the adult.

Perspectives

The recent advances in identifying and counteracting factors that prevent the regeneration of lesioned axons in the adult CNS of higher vertebrates have uncovered an astonishing capacity for local use-dependent plasticity of neuronal circuits in the adult. Such plasticity is apparently responsible for the observation that regeneration of only 1–5% of the axons from a completely lesioned fibre tract, e.g. the cortico-spinal tract essential for voluntary motor control, can be sufficient to induce remarkable restitution of function. Interestingly, functional restitution is detected mainly for tasks that involve simple synaptic circuitry, whereas more complex functions are not or only poorly recovered. Observations made on the restitution of Parkinson-type deficits upon implantation of dopaminergic neurons into the striatum led to similar conclusions. Clearly, therefore, very significant progress towards promoting effective neuro-regeneration in the adult CNS has been made in the last few years. While cautious optimism is no longer a matter of faith, one major challenge still remaining is to develop and refine procedures that would effectively promote axonal regeneration in patients. In addition, however, the recent findings have emphasized the need to understand the factors that control the effectiveness and specificity of structural and functional plasticity in the adult nervous system. The establishment of genetic models in mice and flies promises to provide definitive information about the mechanisms that control axon elongation and use-dependent plasticity. Such information should lead to the identification of promising targets for intervention in patients. Given the dramatic pace at which molecular and systems neuroscience are establishing common areas of research, this promises to be an exciting research field over the next few years.

Summary

- *Specificity of connectivity is essential to nervous system function. It is determined by intrinsic programmes of gene expression that define neuronal phenotypes, and by activity-dependent mechanisms. Neuroregeneration in the adult may involve re-activation of growth programmes within the constraints of neuron-type specific phenotypes.*

- *Lesion-induced re-induction of an axonal growth mode in adult neurons correlates with a vigorous cell body reaction that can also lead to apoptotic cell death. Directing the cell body reaction towards regeneration is a major goal towards improving regeneration.*

- *Extrinsic factors that prevent axonal regeneration in the adult CNS of higher vertebrates include inhibitory components on the surface of oligodendrocytes and CNS myelin, and proteoglycans associated with scar material; grafts of certain glial cells can promote regeneration.*

- *Local nerve sprouting and synaptic plasticity can produce dramatic functional adaptation to lesions in the adult and greatly enhance the impact of the partial regeneration of lesioned axons; nerve sprouting is promoted by diffusible and contact-mediated extrinsic mechanisms, and by intrinsic neuronal components.*

- *As a result of recent discoveries, significant progress in promoting axonal regeneration and recovery of function in the adult can be anticipated.*

References

1. Goodman, C.S. & Shatz, C.J. (1993) Developmental mechanisms that generate precise patterns of neuronal connectivity. *Cell* **72**, 77–98
2. Tanabe, Y. & Jessell, T.M. (1996) Diversity and pattern in the developing spinal cord. *Science* **274**, 1115–1122
3. Davies, A.M. (1994) Intrinsic programmes of growth and survival in developing vertebrate neurons. *Trends Neurosci.* **17**, 195–199
4. Skene, J.H.P. (1989) Axonal growth-associated proteins. *Annu. Rev. Neurosci.* **12**, 127–156
5. Bisby, M.A. & Tetzlaff, W. (1992) Changes in cytoskeletal protein synthesis following axon injury and during regeneration. *Mol. Neurobiol.* **6**, 107–123
6. Caroni, P. (1997) Intrinsic neuronal determinants that promote axonal sprouting and elongation. *BioEssays* **19**, 767–775
7. Darian-Smith, C. & Gilbert, C.D. (1994) Axonal sprouting accompanies functional reorganization in adult cat striate cortex. *Nature (London)* **368**, 737–740
8. Das, A. & Gilbert, C.D. (1995) Long-range horizontal connections and their role in cortical reorganization revealed by optical recording of cat primary visual cortex. *Nature (London)* **375**, 780–784
9. Berkelaar, M., Clarke, D.B., Wang, Y.C., Bray, G.M. & Aguayo, A.J. (1994) Axotomy results in delayed death and apoptosis of retinal ganglion cells in adult rats. *J. Neurosci.* **14**, 4368–4374
10. Herdegen, M., Skene, J.H.P. & Bähr, M. (1997) The c-jun transcription factor – bipotential mediator of neuronal death, survival and regeneration. *Trends Neurosci.* **20**, 227–231

11. Villegas-Perez, M.P., Vidal-Sanz, M., Bray, G.M. & Aguayo, A.J. (1988) Influences of peripheral nerve grafts on the survival and regrowth of axotomized retinal ganglion cells in adult rats. *J. Neurosci.* **8**, 265–280

12. Doster, S.K., Lozano, A.M., Aguayo, A.J. & Willard, M.B. (1991) Expression of the growth-associated protein GAP-43 in adult rat retinal ganglion cells following axon injury. *Neuron* **6**, 635–647

13. McKerracher, L., Essagian, C. & Aguayo, A.J. (1993) Marked increase in beta-tubulin mRNA expression during regeneration of axotomized retinal ganglion cells in adult mammals. *J. Neurosci.* **13**, 5294–5300

14. Richardson, P.M. & Issa, V.M.K. (1984) Peripheral nerve injury enhances central regeneration of primary sensory neurones. *Nature (London)* **309**, 791–793

15. Chong, M.S., Woolf, C.J., Turmaine, M., Emson, P.C. & Anderson, P.N. (1996) Intrinsic versus extrinsic factors in determining the regeneration of the central processes of rat dorsal root ganglion neurons: the influence of a peripheral nerve graft. *J. Comp. Neurol.* **370**, 97–104

16. Benfey, M. & Aguayo, A.J. (1982) Extensive elongation of axons from rat brain into peripheral nerve grafts. *Nature (London)* **296**, 150–152

17. Schwab, M.E. & Bartholdi, D. (1996) Degeneration and regeneration of axons in the lesioned spinal cord. *Physiol. Rev.* **76**, 319–370

18. Rossi, F., Jankovski, A. & Sotelo, C. (1995) Differential regenerative response of Purkinje cell and inferior olivary axons confronted with embryonic grafts: environmental cues versus intrinsic neuronal determinants. *J. Comp. Neurol.* **359**, 663–677

19. Davies, S.J.A., Fitch, M.T., Memberg, S.P., Hall, A.K., Raisman, G. & Silver, J. (1997) Regeneration of adult axons in white matter tracts of the central nervous system. *Nature (London)* **390**, 680–683

20. Li, Y., Field, M. & Raisman, G. (1997) Repair of adult corticospinal tract by transplants of olfactory ensheating cells. *Science* **277**, 2000–2002

21. Wictorin, K., Brundin, P., Gustavii, P., Lindvall, O. & Björklund, A. (1990) Reformation of long axon pathways in adult rat central nervous system by human forebrain neuroblasts. *Nature (London)* **347**, 556–558

22. Chen, D.F., Jahveri, S. & Schneider, G.E. (1995) Intrinsic changes in developing retinal neurons result in regenerative failure of their axons. *Proc. Natl. Acad. Sci. U.S.A.* **92**, 7287–7291

23. Chen, D.F., Schneider, G.E., Martinou, J.C. & Tonegawa, S. (1997) Bcl-2 promotes regeneration of severed axons in mammalian CNS. *Nature (London)* **385**, 434–439

24. Deller, T., Frotscher, M. & Nitsch, R. (1995) Morphological evidence for the sprouting of inhibitory commissural fibers in response to the lesion of the excitatory entorhinal input to the rat dentate gyrus. *J. Neurosci.* **15**, 6868–6878

25. Bailey, C.H., Alberini, C., Ghirardi, M. & Kandel, E.R. (1994) Molecular and structural changes underlying long-term memory storage in *Aplysia*. *Adv. Second Messenger Phosphoprotein Res.* **29**, 529–544

26. Tessier-Lavigne, M. & Goodman, C.S. (1996) The molecular biology of axon guidance. *Science* **274**, 1123–1132

27. Son, Y.-J. & Thompson, W.J. (1995) Nerve sprouting in muscle is induced and guided by processes extended by Schwann cells. *Neuron* **14**, 133–141

28. Aigner, L., Arber, S., Kapfhammer, J.P., Laux, T., Schneider, C., Botteri, F., Brenner, H.-R. & Caroni, P. (1995) Overexpression of the neural growth-associated protein GAP-43 induces nerve sprouting in the adult nervous system of transgenic mice. *Cell* **83**, 269–278

29. Caroni, P., Aigner, L. & Schneider, C. (1997) Intrinsic neuronal determinants locally regulate extrasynaptic and synaptic growth at the adult neuromuscular junction. *J. Cell Biol.* **136**, 679–692

6

A molecular basis for opiate action

Dominique Massotte and Brigitte L. Kieffer[1]

*Département des récepteurs et protéines membranaires
(CNRS UPR 9050), Ecole Supérieure de Biotechnologie de Strasbourg,
F-67400 Illkirch-Graffenstaden, France*

Introduction

Alkaloids, from the Arabian word *al-qali* meaning saltwort, consist of a large family of compounds synthesized almost exclusively in plants. In spite of their elusive biological role, a lot of the molecules that accumulate in the plant cell vacuoles may exert strong effects on animals and humans. Opiates are alkaloids which have been known and used for some 4000 years, since the Sumerians cultivated poppies (*Papaver somniferum*) and isolated the liquid appearing on notched unripe seed capsules. The so-called opium (from the Greek word *opos*, meaning juice) may first have been employed in religious rituals because it evokes euphoria. It was also long used to fight coughing or diarrhoea, or to relieve pain, although variability in both the quality of preparations and the rate of absorbance made its use rather uncertain. Opium seems to have been introduced by the Arabs to India and Asia during the 8th century and to have reached Europe between the 10th and 13th centuries. Addiction resulting from its repeated use was reported as early as the 16th century, but the situation was nowhere as acute as in China, where opium replaced the banned tobacco smoking in the mid-17th century.

In 1806 Sertürner isolated the first active substance from opium, which he named morphine (Figure 1) after the god of dreams, Morpheus. A few years later codeine, an intermediate of the morphine biosynthetic pathway, was isolated, and several other constituents were subsequently identified (e.g. papaverine, noscapine, thebaine). After the invention of the hypodermic syringe and hollow

[1]*To whom correspondence should be addressed.*

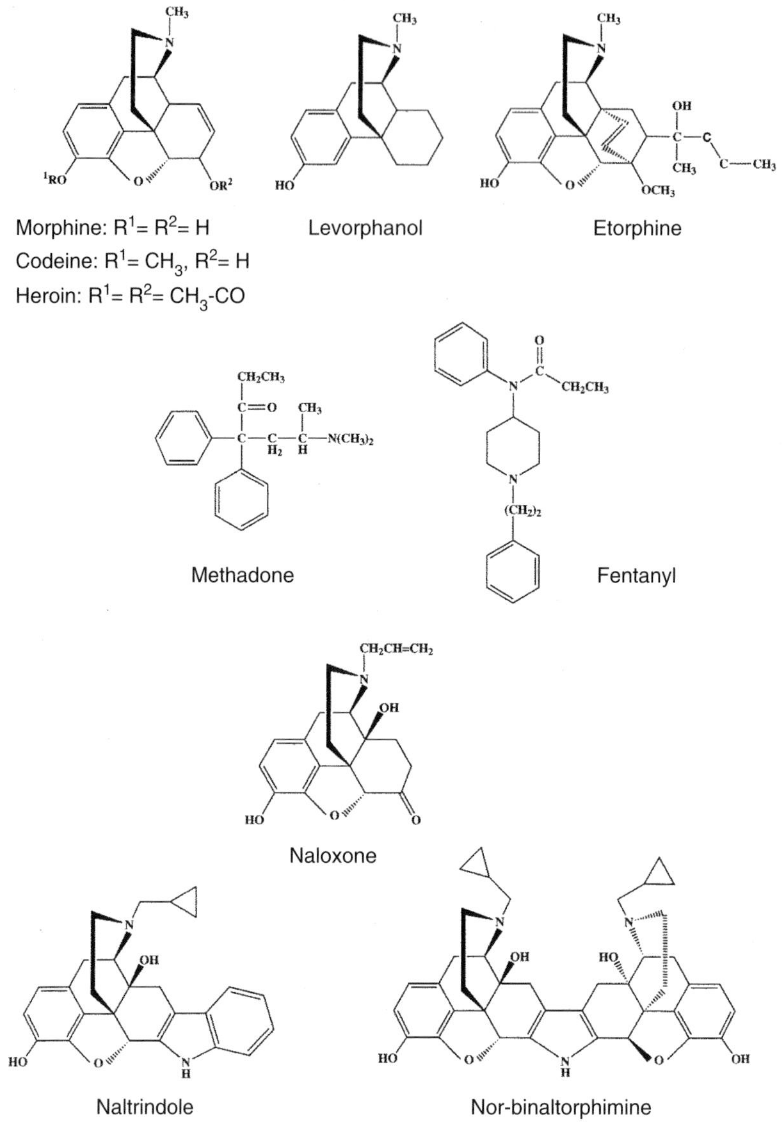

Morphine: R^1= R^2= H
Codeine: R^1= CH$_3$, R^2= H
Heroin: R^1= R^2= CH$_3$-CO

Levorphanol

Etorphine

Methadone

Fentanyl

Naloxone

Naltrindole

Nor-binaltorphimine

Figure 1. Chemical structures of some natural and synthetic alkaloids
Prototypic μ-receptor agonists are shown at the top. Bulky aliphatic N-substituents confer antagonistic properties and other substitutions may render alkaloids highly selective towards a specific receptor subtype. The classical antagonists naloxone (non-selective opioid), naltrindole (δ-selective) and nor-binaltorphimine (κ-selective) are shown at the bottom.

needle in the 1850s, morphine was slowly introduced into medicine for the treatment of severe or chronic pain, as well as in surgical practice to reduce postoperative pain or as an adjunct to general anaesthetics. During the next 100 years many attempts were made to find molecules which could mimic morphine analgesia but which were devoid of its secondary effects (nausea, constipation, respiratory depression) and would not induce addiction. In 1898 heroin (Figure 1) was synthesized with the hope that it would fulfil these requirements. Unfortunately, heroin was later found to be even more addictive than morphine and became a major drug of abuse. Indeed heroin is more lipid-soluble than morphine and can therefore readily enter the brain, where it is converted into morphine and produces the euphoria or 'high' anticipated by drug addicts.

The opioid system: discovery of a complex neurotransmitter system

Several observations suggested that opiates interact with specific binding sites, which were likely to be receptors. This could be shown when radioligands with high specific activities were developed. In 1973, three groups [1–3] showed simultaneously the existence of high-affinity, saturable, stereospecific binding sites for [^{3}H]naloxone or [^{3}H]etorphine on brain membranes. In 1976, Martin et al. [4] reported the first evidence for multiple opioid receptors. Pharmacological studies led to the classification of opioid-binding sites into three receptor types, referred as to μ, δ and κ receptors. Later, the availability of numerous synthetic opiates and their use in biological assays indicated a possible heterogeneity within each receptor class, and the existence of δ_1 and δ_2, μ_1 and μ_2, and κ_{1a} κ_{1b}, κ_2 and κ_3 opioid receptors was postulated [5].

The demonstration of the existence of opioid receptors suggested that the receptor sites might be the target for endogenous opiate-like (named opioid) molecules. In 1975, Hughes et al. [6] isolated two pentapeptides, Leu-enkephalin (YGGFL) and Met-enkephalin (YGGFM), from pig brain. Shortly afterwards, other endogenous peptides were identified (Table 1). Opioid peptides are derived from proteolysis of larger precursor proteins which are encoded by three distinct genes (Figure 2) [7].

Endogenous opioid peptides exhibit weak selectivity towards the three opioid receptor types. β-Endorphin binds to μ and δ receptors with comparable affinity. Met- and Leu-enkephalins are considered to be endogenous ligands for δ receptors, and similarly dynorphins for κ receptors. No endogenous peptide with high affinity and specificity for the μ receptor type was described until Zadina et al. [8] reported the characterization of two novel tetrapeptides isolated from bovine cortex. This discovery resulted from an unusual interplay between pharmacology and combinatorial chemistry. The isolated peptides, endomorphin 1 (YPWF-NH$_2$) and endomorphin 2 (YPFF-NH$_2$), were found to bind to the μ receptor with affinities up to 24 times higher than other known endogenous opioid peptides, and with 4000- and 15000-fold preference over binding to the δ and κ receptors respectively. Definite demonstration of their physiological existence now requires the identification of the encoding genes. Besides the four categories of endogenous opioid peptides described above, a few other natural peptides have been isolated which show opioid activity [9].

The identification of opioid receptors together with their specific endogenous ligands led to the suggestion that they could organize into a complex neurotransmitter system (Figure 3). Indeed, besides a major role in

Table 1. Sequences of major endogenous opioid peptides

Opioid peptides are processed from the three precursor proteins pro-opiomelano-cortin (POMC), pro-enkephalin and pro-dynorphin (see Figure 2). Amino acid sequences are given using the single-letter code. The numbers in parentheses indicate the sequence residues of the cleavage products with opioid activity.

Precursor	Opioid peptides	Amino acid sequence
POMC	β-Endorphin (1–31, 1–27, 1–26)	**YGGF**MTSEKSQTPLVTLFKNAIIKNAYKKGE
	γ-Endorphin [= β-Endorphin-(1–17)]	**YGGF**MTSEKSQTPLVTL
	α-Endorphin [= β-Endorphin-(1–16)]	**YGGF**MTSEKSQTPLVT
Pro-enkephalin	Leu-enkephalin	**YGGF**L
	Met-enkephalin	**YGGF**M
	Heptapeptide	**YGGF**MRF
	Octapeptide	**YGGF**MRGL
Pro-dynorphin	α-Neo-endorphin	**YGGF**LRKYPK
	β-Neo-endorphin	**YGGF**LRKYP
	Dynorphin A (1–32, 1–17, 1–13, 1–8)	**YGGF**LRRIRPKLKWNNQKRYGGFLRRQFKVVT
	Dynorphin B (1–29, 1–13)	**YGGF**LRRQFKVVTRSQEDPNAYSGELFDA
	Leu-enkephalin	**YGGF**L

endogenous pain-controlling pathways (see [10]), the opioid system is involved in a large variety of biological events (reviewed annually [11]). Numerous pharmacological studies have demonstrated that the opioid system plays a role in stress-induced analgesia [7] and contributes to some pathophysiological conditions associated with stress. Regulation of affective behaviour, including motivation and reward, is another crucial function of the opioid system. Indeed the dysregulation of rewarding pathways by exogenous opiates is thought to underlie opiate addiction by neurobiological [12] and molecular [13] mechanisms which are being intensively investigated. Opioids might also play a role in cognitive functions, such as learning and memory, and alterations in opioidergic systems have been hypothesized to be associated with some psychiatric or neurological disorders. The opioid system modulates neuroendocrine physiology and controls autonomic functions, such as respiration, blood pressure, thermoregulation and gastrointestinal motility, and is involved in immune function.

Anatomical studies have shown that the components of the opioid system are widely distributed throughout the central nervous system (CNS) [14], in agreement with the broad diversity of opiate biological actions. Noteworthy is the

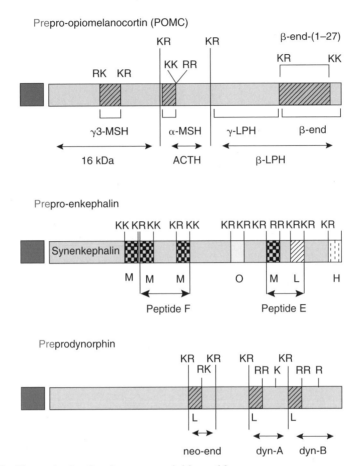

Figure 2. Biosynthesis of endogenous opioid peptides

Opioid peptides are derived from three different precursor proteins, from which the signal sequence (dark blue box) is cleaved by a signal peptidase. The three precursor proteins are processed at paired basic residues [Lys (K) or Arg (R)] and give rise to a number of biologically active peptides, which may be processed further into smaller active products (see Table 1 for opioid peptides). POMC contains a single β-endorphin sequence at its C-terminus (β-end). This precursor produces γ_3-melanocyte-stimulating hormone (γ3-MSH), adrenocorticotropin (ACTH) and β-lipotropin, the latter two being processed further into α-melanocyte-stimulating hormone (α-MSH) and γ-lipotropin (γ-LPH) + β-endorphin respectively. Cleavage of pro-enkephalin gives rise to the non-opioid peptide synenkephalin, and four copies of Met-enkephalin (M), one copy of Leu-enkephalin (L), one copy of the heptapeptide Met-enkephalin-Arg[6]-Phe[7] (H) and one copy of the octapeptide Met-enkephalin-Arg[6]-Gly[7]-Leu[8] (O). Larger products can also be detected, such as peptide F and peptide E. Pro-dynorphin processing results in α and β neo-endorphins (neo-end), dynorphin A (dyn-A) and dynorphin B (dyn-B), with all three sequences including a copy of Leu-enkephalin (L).

observation that μ, δ and κ binding sites display distinct distribution patterns, indicating that each receptor class contributes differently to opioid funtion. On the peptide side, prepro-opiomelanocortin (POMC) synthesis is highly restricted, while prepro-enkephalin and prepro-dynorphin transcripts both display a similar

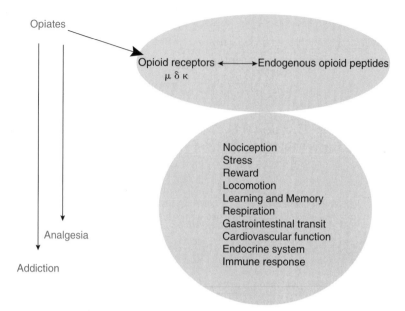

Figure 3. The complex opioid neurotransmitter system
The endogenous opioid system (blue oval) consists of three classes of opioid receptors and a family of opioid peptides which interact to modulate a wide variety of physiological functions (blue circle). Opioid receptors are also the target for natural exogenous substances (opiates) and synthetic analogues which induce analgesia or may lead to drug addiction.

widespread expression pattern in the CNS. Some reports indicate that both receptor sites and peptides are also present in non-neuronal tissues.

In summary, opioids act generally along neural pathways related to behaviour that is essential for self and species survival, and it is widely accepted that the endogenous opioid network is recruited in response to threatening stimuli.

Opioid receptors: first steps towards molecular mechanisms of opioid action

In 1992, some 20 years after opioid receptors had been postulated, a cDNA encoding the mouse δ-opioid receptor (*mDOR*) was isolated simultaneously by two independent laboratories using expression cloning in mammalian cells [15,16]. Both groups constructed a random-primed expression library from the rat/mouse glioma × neuroblastoma hybrid cell line (NG108-15) known to express the δ-opioid receptor at high levels. The library was transfected into a mammalian cell line (COS cells) lacking endogenous opioid receptors. Cells expressing the opioid receptor were detected using a pharmacological assay relying on high-affinity binding of a δ-receptor-specific radiolabelled ligand, and both groups isolated the same cDNA using enkephalin-derived peptides as probes. Cloning by homology was then initiated using *mDOR* as a probe in PCR or low-stringency screening procedures, leading to the identification of cDNAs encoding δ (*DOR*), μ (*MOR*) and κ (*KOR*) opioid receptors in rodents and humans [17,18]. Genes encoding opioid receptors have now been identified throughout vertebrate evolution.

Binding studies using recombinant mouse, rat and human receptors expressed in heterologous host cells unambiguously assigned each of the three cloned receptors to the three opioid receptor classes described previously in nervous tissues. Interestingly, only one gene has been isolated for each opioid receptor type. Genes for the multiple μ (μ_1, μ_2), δ (δ_1, δ_2) and κ (κ_1, κ_2, κ_3) receptor subtypes have not been isolated by molecular approaches, nor have alternative splicing mechanisms been identified that could be responsible for the postulated receptor heterogeneity. Perhaps other opioid receptor subtypes are encoded by genes structurally unrelated to the known *MOR*, *DOR* or *KOR* genes. Alternatively, receptor variants could arise from distinct conformations of the three known receptor proteins depending on the cellular context or intrinsic properties of the interacting ligand.

Analysis of the cloned opioid receptor sequences indicated that they belong to the G-protein-coupled receptor family (Figure 4). The three opioid receptor types show high sequence similarity with somatostatin, angiotensin and chemoattractant receptors. Seven putative transmembrane α-helical segments were identified, as well as potential glycosylation sites in the N-terminal domain and several phosphorylation sites in the third intracellular loop and the

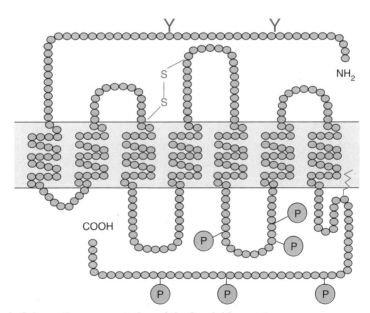

Figure 4. Schematic representation of the δ-opioid receptor
The μ, δ, and κ receptor subtypes are highly similar. Each circle indicates an amino acid residue, and the seven putative α-helical segments are represented inserted in the membrane. Putative glycosylation (Y), phosphorylation (P) and palmitoylation (blue zig-zag line) sites and a postulated disulphide bridge are represented. The N-terminal part of the protein is located on the extracellular side of the membrane and the C-terminal part on the cytoplasmic face. The deduced protein sequences of all the cloned receptors (μ, δ, κ of mouse, rat and human origin) are given in [17].

C-terminus. These consensus sites for post-translational modifications are located at different positions in the *MOR*, *DOR* and *KOR* sequences. By analogy with other receptors, the presence of an important disulphide bridge between the first and second extracellular loops was also postulated.

The cloning of the opioid receptor genes opened up a new era in the molecular understanding of the structure–activity relationships in opioid receptors. It is now possible to express the receptors as recombinant proteins in various cell lines, to modify their sequences by site-directed mutagenesis or to create chimaeras by fusing parts of different receptor types. Ligand recognition, signal transduction and receptor desensitization can now be analysed extensively using unlimited sources of proteins. Furthermore, structure–function studies may now be performed on human recombinant receptors, an important factor in drug screening programmes.

Ligand recognition is presently being explored by many groups, and several three-dimensional models have been proposed (reviewed in [19]). Mutagenesis data show that extracellular loops, which differ greatly between opioid receptors, are critical for μ, δ and κ selectivity. On the other hand, the three opioid receptor types share high sequence homology in the transmembrane regions, and an opioid-binding pocket has been located within the helical bundle. Accordingly, studies of point mutants showed that a number of transmembrane residues found in helices 2–7, and conserved across opioid receptor types, are important to maintain ligand binding. Interestingly, descriptions of ligand-binding mechanisms led to the conclusion that there is no unique opioid-binding pocket, but rather a specific network of multiple interactions within each ligand–receptor complex [20]. More extensive modelling studies and experimental identification of critical structural determinants of the receptors will be available once the three-dimensional structure is resolved.

Biochemical and electrophysiological studies on various primary neuron cultures and neuroblastoma cell lines have shown that the μ, δ and κ receptors inhibit the cAMP pathway, decrease the conductance of various voltage-gated Ca^{2+} channels or activate inwardly rectifying K^+ channels, depending on the cell under study (Figure 5). Pertussis-sensitive $G\alpha_i$ or $G\alpha_o$ types seem to be the preferred coupling partners of opioid receptors over other $G\alpha$ protein types. Functional coupling to cellular effectors has now been demonstrated for the recombinant receptors produced in a number of heterologous expression systems [17]. Such studies have also shown that not only the intrinsic properties of the receptor, but also the availability of G-proteins and downstream effectors in the cell, drive the signalling process. As an example, co-expression of a $G\alpha$ protein subunit modifies receptor–G-protein interactions and switches the receptor coupling from the endogenous to the heterologous α subunit. Functional studies have also demonstrated that each receptor type may simultaneously activate multiple G-protein α subunits, as well as multiple effector pathways in the same cell. Peptide competition studies showed that the second

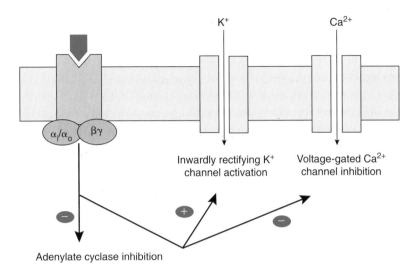

Figure 5. Functional coupling of opioid receptors
Upon activation by an agonist (dark blue arrow), all three opioid receptor subtypes can inhibit adenylate cyclase activity and voltage-gated Ca^{2+} channels or activate inwardly rectifying K^+ channels. These effects are mediated by heterotrimeric G-proteins whose $G\alpha$ subunit is of the G_i or G_o type. Coupling to these effector systems has been demonstrated in nervous tissue and using heterologous expression in various host cells.

and third intracellular loops, as well as the membrane-proximal part of the C-terminal tail, interact with G-proteins.

Signal transduction is tightly associated with receptor desensitization, defined as a rapid loss of receptor function upon sustained exposure to an agonist. This process tends to limit the biological action of the stimulating drug or neurotransmitter at the cellular level. Opioid receptor desensitization was first shown in neuroblastoma cell lines expressing endogenous receptors or in isolated organ preparations, and has now been described in several heterologous host cell lines expressing the recombinant receptors. Such model cell lines provide convenient tools with which to dissect the molecular events underlying decreased cellular response under chronic opiate stimulation, including phosphorylation of the receptor by endogenous specific (G-coupled-receptor kinases) or non-specific (protein kinase A or C) kinases, receptor internalization and down-regulation [19]. These studies represent a first step towards the elucidation of opiate tolerance and dependence at the cellular level [21]. The situation *in vivo* is much more complex, and the chronic activation of opioid receptors triggers long-term adaptative changes at the level of neuronal networks [12,13] that ultimately lead to opiate addiction. Although these phenomena are highly complex and poorly understood, future studies should clarify whether opioid receptor desensitization plays a role in the development of opiate tolerance and dependence *in vivo*.

Recent approaches: from gene to function

Gene targeting techniques allow us to inactivate any gene of interest in mice. This is an extremely powerful approach to investigating the action of a defined gene product in development, as well as in any physiological function or behaviour in the adult. Genes encoding all the components of the opioid system (opioid peptides and receptors) have now been disrupted in mice by homologous recombination [22]. All these mutant mice reached the adult stage without apparent anatomical abnormalities, indicating that the absence of a single component of the opioid system is not detrimental to development. Although only a few behavioural studies have been performed to date, the first results have indicated increased pain thresholds in mice lacking prepro-enkephalin, μ or κ receptors, increased anxiety in the prepro-enkephalin knock-out mice and an altered response to stress in β-endorphin-deficient mutant mice. Comparative analysis of the various mutant mice in identical behavioural paradigms will highlight the specific contribution of each component of the endogenous opioid system in the response to stressful situations. The inactivation of multiple opioid peptide and/or receptor genes in the same animal has not yet been described.

Of great interest is the study of opiate action in opioid receptor-deficient mice. These mutant mouse strains represent exquisite tools with which to define the mode of action *in vivo* of prototypic, as well as novel, opioid compounds at the molecular level. Morphine was long believed to act at multiple receptors *in vivo*, because morphine affinity towards μ receptor binding sites exceeds that towards δ and κ receptors by only 100-fold. Although this selectivity factor is reasonably high for *in vitro* experiments, where the amount of drug available to the receptors is well controlled, it is rather low when considering *in vivo* protocols using repeated systemic administration of high doses of drug. The pharmacological action of morphine was extensively investigated by Matthes et al. [23] in mice lacking the μ-opioid receptor. These experiments showed clearly that all responses to morphine were abolished in mutant mice, whatever the dose or mode of administration. Indeed, morphine was unable to induce analgesia, reward or physical dependence in mice lacking the μ receptors [23]. Furthermore, both respiratory depression and immunosuppression, well-established morphine side-effects, were absent in these mice. The genetic approach, therefore, nicely demonstrated that the receptor protein encoded by the *MOR* gene represents the essential molecular target of morphine *in vivo*. As a consequence, it seems that both the desired effects and the adverse side-effects of morphine, one of the most clinically useful opiates, result from the activation of the same receptor protein. Therefore it appears unlikely that an ideal analgesic could be developed by drug design strategies that adopt the *MOR*-encoded receptor as their target.

Perspectives: an arduous march to therapeutics

Numerous morphine derivatives have been synthesized with the hope that they would have analgesic effects comparable with those of morphine but without its secondary effects of respiratory depression, constipation, abuse liability, euphoria, nausea and vomiting. The hope for new efficient analgesics, free of the usual morphine side-effects, has motivated the synthesis of many new molecules. Numerous prototypic ligands, peptidic analogues and non-peptide opioid alkaloid derivatives have been developed for each of the opioid receptors, both as pharmacological tools and as potential therapeutic agents [24]. These ligands are characterized by high-affinity binding and significant μ, δ and κ selectivity, as well as by greater stability in the case of the peptide derivatives. Most of these compounds contain the classical opioid pharmacophore, i.e. a phenolic ring located at a fixed distance from a positively charged nitrogen atom, a structure which is common to opioid peptides and morphine analogues.

A tremendous diversity of compounds has been studied in receptor binding experiments, isolated organ bioassays and physiological studies *in vivo*. None of the synthetic molecules has yet proven entirely satisfactory; however, some of these compounds are now used as therapeutic agents. The development of piperidine derivatives, such as Fentanyl (Figure 1), was pionereed by Janssen and led to the synthesis of several potent analgesics with different clinical applications. The thebaine-derived etorphine is about 1000 times more potent than morphine, but its side-effects restrict its veterinary use as a sedative for large animals. Morphinans, e.g. levorphanol (Figure 1), were created by removal of one or more of the morphine rings. Compounds of the methadone type represent the ultimate simplification of the morphine structure. One of these is a widely used analgesic (*d*-propoxyphene, or Darvon), while methadone itself (Figure 1), a μ-receptor partial agonist, is used to withdraw human addicts because it can be taken orally, has long-lasting action and does not induce detectable psychotropic effects.

All opiate drugs used clinically act mainly at the μ-receptor. κ-Agonists have been developed with the hope that their opposing action in mood control compared with that of μ- and δ-agonists would make them non-addictive. Unfortunately, their strong aversive properties in humans hamper their use for therapeutic purposes. Great efforts are now being made to design highly selective and potent δ-agonist compounds, to develop peripherally acting opiates that would be devoid of central effects, and to combine low doses of opiates with drugs acting at different pharmacological targets.

In conclusion, the cloning of opioid receptor genes has recently opened up an entirely new field of investigation in opioid research. The amount of information obtained from these newly developed molecular tools should increase exponentially in the very near future. Understanding the molecular basis of opiate action both at cellular level and *in vivo* will allow us to establish novel therapeutic strategies for the treatment of pain and addiction.

Summary

* *Opioid receptors mediate the strong analgesic and addictive actions of exogenous opiates, the prototype of which is morphine.*
* *The opioid system consists of a family of endogenous opioid peptides and three receptor types, m, d and k. It is widely distributed through-out the CNS and regulates a large diversity of physiological functions, including pain perception and mood control.*
* *The recent cloning of opioid receptors has opened up a new era in opi-oid research. The molecular basis of opioid action may now be addressed by in vitro structure–function studies of recombinant recep-tors and by in vivo gene targeting.*
* *Novel drug design strategies based on data obtained from molecular approaches will be developed in order to generate the long-sought non-addictive analgesic.*

We thank Dr. Franc Pattus and Professor Pierre Chambon for constant support, and Dr. K. Befort for critical reading of the manuscript.

References

1. Pert, C.B. & Snyder, S.H. (1973) Opiate receptor: demonstration in nervous tissue. *Science* **179**, 1011–1014
2. Simon, E.J., Hiller, J.M. & Edelman, I. (1973) Stereospecific binding of the potent narcotic analgesic ³H etorphine to rat brain homogenate. *Proc. Natl. Acad. Sci. U.S.A.* **70**, 1947–1949
3. Terenius, L. (1973) Stereospecific interaction between narcotic analgesics and a synaptic plasma membrane fraction of rat cerebral cortex. *Acta Pharmacol. Toxicol.* **32**, 317–319
4. Martin, W.R., Eades, C.G., Thompson, J.A., Huppler, R.E. & Gilbert, P.E. (1976) The effects of morphine- and nalorphine-like drugs in the non-dependent and morphine-dependent chronic spinal dog. *J.Pharmacol. Exp. Ther.* **197**, 517–532
5. Pasternak, G.W. (1993) Pharmacological mechanisms of opioid analgesics. *Clin. Neuropharmacol.* **16**, 1–18
6. Hughes, J., Smith, T.W., Kosterlitz, L.A., Fothergill, L.A., Morgan, B.A. & Morris, H.R. (1975) Identification of two related pentapeptides from the brain with potent opiate agonist activity. *Nature (London)* **258**, 577–579
7. Akil, H., Watson, S.T., Young, E., Lewis, M.E., Khachaturian, H. & Walker, J.M. (1984) Endogenous opioids: biology and function. *Annu. Rev. Neurosci.* **7**, 223–255
8. Zadina, J.E., Hackler, L., Ge, L.-J. & Kastin, A.J. (1997) A potent and selective endogenous agonist for the μ-opiate receptor. *Nature (London)* **386**, 499–502
9. Teschemacher, H. (1993) Atypical opioid peptides. *Handb. Exp. Pharmacol.: Opioids I* (Hertz, A., ed.) pp. 499–528, Springer Verlag, Berlin
10. Dickenson, A.H. (1991) Mechanisms of the analgesic actions of opiates and opioids. *Br. Med. Bull.* **47**, 690–702
11. Olson, G.A. (1996) Endogenous opiates. *Peptides (NY)* **17**, 1421–1466
12. Koob, G.F. & Le Moal, M. (1997) Drug abuse: hedonic homeostatic dysregulation. *Science* **278**, 52–58
13. Nestler, E.J. & Aghajanian, G.K. (1997) Molecular and cellular basis of addiction. *Science* **278**, 58–63

14. Khachaturian, H., Schaefer, M.K.H. & Lewis, M.E. (1993) Anatomy and function of the endoge-nous opioid systems. *Handb. Exp. Pharmacol.: Opioids I* (Hertz, A., ed.) pp. 471–497, Springer Verlag, Berlin

15. Kieffer, B.L., Befort, K., Gaveriaux-Ruff, C. & Hirth, C.G. (1992) The δ-opioid receptor: isolation of a cDNA by expression cloning and pharmacological characterization. *Proc. Natl. Acad. Sci. U.S.A.* **89**, 12048–12052

16. Evans, C.J., Keith, D.E., Morrison, H., Magendzo, K. & Edwards, R.H. (1992) Cloning of a δ-opioid receptor by functional expression. *Science* **258**, 1952–1955

17. Kieffer, B.L. (1997) Molecular aspect of opioid receptors. *Handb. Exp. Pharmacol.: The Pharmacology of Pain* (Dickenson, A. and Besson, J.M., eds.), pp. 281–303, Springer Verlag, Berlin

18. Satoh, M. & Minami, M. (1995) Molecular pharmacology of the opioid receptors. *Pharmacol. Ther.* **68**, 343–364

19. Befort, K. & Kieffer, B.L. (1997) Structure–activity relationships in the δ-opioid receptor. *Pain Rev.* **4**, 100–121

20. Befort, K., Tabbara, L., Kling, D., Maigret, B. & Kieffer, B.L. (1996) Role of aromatic transmem-brane residues of the δ-opioid receptor in ligand recognition. *J. Biol. Chem.* **271**, 10161–10168

21. Cox, B.M. (1993) Opioid receptor–G protein interactions: acute and chronic effects of opioids. *Handb. Exp. Pharmacol.: Opioids I* (Hertz, A., ed.), pp. 145–188, Springer Verlag, Berlin

22. Gaveriaux-Ruff, C. & Kieffer, B.L. (1998) Opioid receptors, gene structure and function. In *Opioids in Pain Control: Basic and Clinical Aspects* (Stein, C., ed.), in the press

23. Matthes, H.W.D., Maldonado, R., Simonin, F., Valverde, O., Slowe, S., Kitchen, I., Befort, K., Dierich, A., Le Meur, M., Dollé, P., et al. (1996) Loss of morphine-induced analgesia, reward effect and withdrawal symptoms in mice lacking the μ-opioid-receptor gene. *Nature (London)* **383**, 819–823

24. Dhawan, B.N., Cesselin, F., Raghubir, R., Reisine, T., Bradley, P.B., Portoghese, P.S. & Hamon, M. (1996) International Union of Pharmacology XII: classification of opioid receptors. *Pharmacol. Rev.* **48**, 567–591

7

Gases as neurotransmitters

Jane E. Haley

Department of Pharmacology, University College London, Gower Street, London WC1E 6BT, U.K.

Introduction

A decade ago, who would have believed that two highly toxic, environmentally polluting gases found in the exhaust emissions of cars could be candidates as important signalling molecules in the body? Nevertheless, we reach the late 1990s in just such a position, with nitric oxide (NO) and carbon monoxide (CO) being implicated in many cellular functions, ranging from control of vascular tone and penile erection to mediating pain and learning. Biologists first became aware of this new class of gaseous mediators in 1987, when a vasodilator produced by endothelial cells, termed 'endothelium-derived relaxing factor' (EDRF) [1], was identified as being the very simple molecule NO [2]. The following year Garthwaite and colleagues created an entirely new area of research when they suggested that, far from being restricted to the vasculature, EDRF (NO) may also be generated in the central nervous system (CNS) and mediate glutamate receptor stimulation of cGMP production in the cerebellum [3]. Following these two discoveries, the number of scientific publications on NO rose dramatically from less than 100 in 1987 and 1988 to 4304 in 1996! NO has now been implicated in many processes, including smooth muscle relaxation following activation of peripheral non-adrenergic non-cholinergic nerves innervating the anococcygeus and retractor penis muscles, where NO or an NO-like molecule may be a neurotransmitter [4]. NO may also be involved in macrophage cytotoxicity, synaptic plasticity in the CNS and neurotoxicity. It is clear, therefore, that this review cannot possibly cover all aspects of NO and CO function. Instead I will focus on just two areas of CNS research where NO and CO have been investigated: in the generation of long-term potentiation (LTP) in the hippocampus, a cellular model for some types of learning, and in the processing of nociceptive (pain) signals in the spinal cord.

Why is everyone so interested in these gases?

Although gaseous in the atmosphere, these molecules do not exist as gases within the body, but are instead dissolved within the cell cytosol or extracellular fluid. Being readily soluble but also highly lipophilic, these molecules can pass easily through the cell membrane following their synthesis, which occurs only when they are required. They therefore differ fundamentally from classical neurotransmitters, which are not membrane permeable, are synthesized in advance and are stored in vesicles awaiting release following calcium influx into the presynaptic terminal. This ability to diffuse out of cells allows NO and CO to be 'released' from any part of the cell (e.g. from a postsynaptic cell dendrite or cell body), whereas traditional neurotransmitters require the vesicular release machinery present in nerve terminals. In addition, NO and CO can also access the intracellular environment of nearby cells by membrane diffusion, interacting directly with enzymes without needing an extracellular receptor. This provides a flexibility that is not achieved with classical neurotransmitters, and 'transmission' can occur in non-classical directions, e.g. from postsynaptic cell to presynaptic cell or from postsynaptic cell to neighbouring postsynaptic cells. NO has the additional property of being short-lived *in vivo* (the free radical is readily oxidized) with a half-life of about 5 s, thus restricting the region of diffusion and conferring some selectivity of action. CO, however, is much more stable and could, in theory, diffuse greater distances.

NO and CO are formed by enzymes

Structure, regulation and activation of NO synthase (NOS) and haem oxygenase (HO)-2

NOS, which generates NO and citrulline from L-arginine, was cloned in 1991 and is now known to consist of three main isoforms: endothelial (eNOS), neuronal (nNOS) and inducible (iNOS) (Figure 1) [5]. NOS is closely related to cytochrome *P*-450 reductase and appears to be an elongated form of this enzyme. It has the same cofactor-binding sites, but contains some additional binding domains, one of which is for the calcium-binding protein calmodulin (Figure 1). This confers calcium sensitivity to eNOS and nNOS, immediately raising the possibility that they may participate in neuronal signalling. iNOS also contains a calmodulin-binding domain, but is activated in a calcium-insensitive manner. All the NOS isoforms also contain a haem-binding domain near the N-terminus and this could have important implications for inhibitor selectivity, as I shall discuss below. Both nNOS and eNOS contain phosphorylation sites and have the potential to be regulated by calcium/calmodulin-dependent kinase II and protein kinases A, C and G. eNOS has an N-terminal

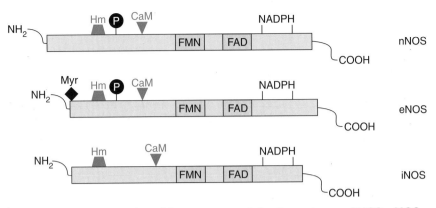

Figure 1. Schematic drawing of the structures of the three classes of NOS: nNOS, eNOS and iNOS
All three isoenzymes contain binding sites for the calcium-sensitive protein calmodulin (CaM) and for haem (Hm). They require NADPH, and bind FMN and FAD tightly. eNOS is the only isoform that contains a myristoylation site (Myr), and this results in association of the enzyme with the plasma membrane. Both nNOS and eNOS contain phosphorylation sites for protein kinase A (circled P), and nNOS activity can be regulated by calcium/calmodulin-dependent kinase II and protein kinases A and G [5].

myristoylation site which is absent from the other isoforms (Figure 1) [5]. Myristoylation sites are found in some proteins that are associated with the plasma membrane, such as G-proteins, and it is thought that myristoylation is required for membrane association. It seems likely, therefore, that eNOS is, or can be, associated with membranes. Indeed, a recent study involving the expression of myristoylation-deficient forms of eNOS seems to confirm this interpretation and suggests that membrane association is required for eNOS to participate in cellular mechanisms [6]. nNOS may also be associated with plasma membrane proteins, but via a different mechanism, as it is able to interact with a postsynaptic density protein (PSD93) which can also interact with the N-methyl-D-aspartate (NMDA) glutamate receptor [7]. It is possible, therefore, that PSD93 could interact simultaneously with nNOS and the NMDA receptor to form large postsynaptic complexes in which the calcium source (NMDA receptor) is coupled tightly to the calcium-sensitive enzyme (nNOS) (see later). iNOS was cloned originally from macrophages and is fundamentally different from eNOS and nNOS as, following cellular stimulation, the enzyme has to be synthesized before it can be activated [5].

Inhibitors of NOS generally act as false substrates, competing for the L-arginine-binding site. While they are specific for NOS, there is little specificity between the various NOS isoforms [8]. A new generation of inhibitors has recently been developed that are derivatives of indazole and are thought to interact with the haem group on NOS. Although these inhibitors are also poorly selective between nNOS and eNOS, one has proved to be a very useful research tool. Although 7-nitroindazole inhibits both eNOS and nNOS *in vitro*, it is unable to inhibit eNOS in intact endothelial cells (how this occurs is poorly understood) and does not, therefore, elevate blood pressure in intact animals, unlike most NOS inhibitors [8]. The selectivity of inhibitors for iNOS is rather better, and several inhibitors are available that do not interact with eNOS/nNOS [8].

CO is a byproduct of the breakdown of haem to biliverdin, catalysed by HO. As with NOS, there is an inducible form of HO (HO-1) and a constitutive form (HO-2), both of which contain a haem-binding pocket [9]. The problem with postulating CO involvement in neuronal function is the apparent lack of an activation signal. NOS contains a calmodulin-binding domain which confers calcium sensitivity to the enzyme, and it is therefore easy to see how NOS may become involved in synaptic modulation. HO-2, however, does not possess such a site; this issue has been largely ignored and really needs to be addressed. Inhibitors of HO also act as false substrates, mimicking the natural substrate haem and competing at the haem-binding site. The inhibitors of HO are not very selective, being able to interact with and inhibit other enzymes containing haem-binding sites, such as NOS and guanylate cyclase (GC) [9].

Localization of NOS and HO-2 within the CNS

Despite the nomenclature used, eNOS is not restricted to endothelial cells, or even to the periphery, and there is now evidence that it can exist in neurons within the CNS, often in the same cell populations as nNOS. Both eNOS and nNOS are present in the olfactory bulb, caudate–putamen, supra-optic nucleus and cerebellum. In the hippocampus eNOS is found in pyramidal neurons, while nNOS is present predominately in interneurons (where its function is not known); it is also present in pyramidal cells, although less abundantly than eNOS [10,11]. Within the spinal cord nNOS is present in the superficial layers of the dorsal horn, where the afferent sensory fibres terminate, and around the central canal in lamina X, which also receives some afferent input [12].

The HO-1 isoform is less abundant in the brain than HO-2, which has a widespread distribution and is present in abundance within the olfactory bulb, cerebellum, brainstem and pyramidal neurons of the hippocampus [9]. As so often happens, the spinal cord has not been included in studies examining the distribution of HO-2 in the brain, and it is not known whether this enzyme is present within the dorsal horn.

What are the targets for NO and CO?

The targets for NO and CO within the CNS are probably very similar, as both molecules can interact with haem moieties. Many enzymes contain haem-binding domains and can become targets for CO and NO. Most extensively studied is GC, the activation of which results in cGMP production. Both NO and CO are activators of GC, although NO is more effective than CO [9]. cGMP has many actions in the CNS: it can open ion channels in the retina, and can activate a cGMP-dependent protein kinase (protein kinase G) and a cGMP-stimulated cyclic nucleotide phosphodiesterase (which hydrolyses cAMP and cGMP) [3]. NO can also activate other enzymes, such as cyclooxygenase (which also contains a haem moiety), and can stimulate auto-ADP-ribosylation of glyceraldehyde-3-phosphate dehydrogenase [3,13]. In addition to activating enzymes, NO (and CO) can also inhibit NOS activity by binding at the haem domain (Figure 1), providing (at least in the case of NO) a negative-feedback system. Apart from its effects on enzymes, NO can also inhibit the NMDA receptor by interacting with an extracellular redox site [13]. Given that NOS and HO-2 are found in abundance in the CNS, it seems likely that more NO and CO targets are waiting to be discovered.

LTP in the hippocampus

Why a diffusible messenger is postulated

LTP is a phenomenon that occurs at certain glutamatergic synapses in the CNS, including the CA3 to CA1 synapse in the hippocampus (Figure 2), and is a cellular model for learning and memory. It is induced by high-frequency tetanic stimulation, and also by depolarization of the postsynaptic cell coupled with low-frequency stimulation of the presynaptic fibres ('paired depolarization'). Following tetanus or paired depolarization there is an increase in the postsynaptic response of the cell. This is observed as an increase in both the amplitude and the speed of depolarization of the excitatory postsynaptic potential (Figure 3), which can last for hours [13]. LTP could result from either an increase in sensitivity of the postsynaptic cell to the neurotransmitter or an increase in neurotransmitter release from the presynaptic cell. There has been controversy over whether a presynaptic or a postsynaptic mechanism is involved in LTP at the CA3 to CA1 pyramidal cell synapse in the hippocampus (the most extensively studied synapse in LTP). Many groups, however, now accept that presynaptic changes can contribute to LTP at this synapse. This leads to a theoretical problem in understanding how LTP is generated, as the activation of postsynaptic NMDA glutamate receptors, and subsequent influx of calcium through this ligand-gated channel, is an absolute requirement for LTP induction at this synapse. How, then, does the presynaptic nerve terminal 'know' when NMDA receptors have been activated on the postsynaptic cell? One explanation is that a 'retrograde messenger' is released from the postsynaptic cell and diffuses to the presynaptic nerve terminal. Until recently, it was thought the postsynaptic side did not possess release machinery. New evidence challenges this assumption, however, and suggests that exocytosis mechanisms may exist at postsynaptic sites after all. This mechanism may not result in transmitter release, however, but could instead be responsible for inserting new receptors into the postsynaptic membrane [14]. It is necessary, therefore, that any retrograde messenger must be able to diffuse through the plasma membrane and across to the presynaptic cell. There are a very small

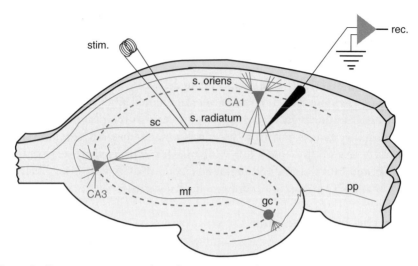

Figure 2. Cartoon representation of a transverse hippocampal slice
Axons of the perforant path (pp) synapse on to the granule cells (gc) of the dentate gyrus. Granule cell mossy fibre axons (mf) synapse on to CA3 pyramidal neurons of the hippocampus, which in turn put out Schaffer collateral axons (sc) that synapse on to both ipsilateral and contralateral CA1 pyramidal neurons in the stratum (s.) radiatum and stratum oriens. The most extensively studied pathway for LTP is the CA3 to CA1 pyramidal synapse, where the Schaffer collateral fibres are stimulated (stim.) and responses are recorded in the CA1 neuron apical dendrites (rec.) in stratum radiatum. The thick broken lines represent the cell body layers of the hippocampus and the dentate gyrus.

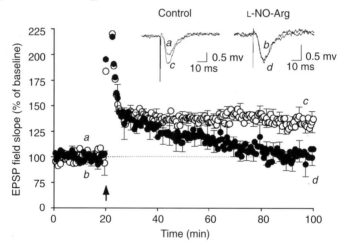

Figure 3. LTP induction is prevented in the presence of NOS inhibitors
Tetanic stimulation (at arrow) of the CA3 to CA1 synapse results in an increase in the rising slope of the recorded field excitatory postsynaptic potential (EPSP; ○), which can last for several hours. Following tetanic stimulation in the presence of an NOS inhibitor (nitro-l-arginine; L-NO-Arg), there is an initial increase in the EPSP slope, but this is not sustained (●). Representative waveforms are displayed above; traces a and b were taken before the tetanus, and c and d were taken following tetanus, in the absence and presence respectively of the NOS inhibitor. Adapted, with permission, from Haley, J.E., Schiable, E., Pavlidis, P., Murdock, A. & Madison, D.V. (1998) Basal and apical synapses of CA1 pyramidal cells employ different LTP induction mechanisms. *Learn. Mem.* **3**, 289–295. ©1998 Cold Spring Harbor Laboratory Press.

number of possible retrograde messengers (i.e. that have the necessary physical properties discussed above); these include the gaseous molecules NO and CO.

NO may be the retrograde messenger

Using inhibitors of NO and CO production, it was immediately clear that NO and CO are not actually 'neurotransmitters' in the hippocampus, as the inhibitors did not alter normal synaptic transmission. NO may, however, have a modulatory role in the CNS, and there is general agreement (although with a few notable exceptions) that NOS inhibitors prevent the induction of LTP in CA1 neurons following either tetanic stimulation or paired depolarization of CA3 axons [13] (Figures 2 and 3). Once LTP has been established, however, NOS inhibitors are without effect, suggesting that, although NO may participate in LTP induction, it is not involved in the maintenance of LTP. Haemoglobin also blocks LTP induction in these cells, but must do so by binding NO or CO extracellularly, as it is so large that it cannot access the intracellular environment [13]. Once NO or CO has left the postsynaptic cell it could act at nearby presynaptic terminals or postsynaptic neurons. Evidence from dual recordings of hippocampal CA1 neurons suggests that such lateral diffusion does indeed occur, as paired depolarization of one neuron (leading to LTP) results in potentiation in a non-paired neighbouring postsynaptic neuron provided that they are close together (150 μm), an effect that is dependent on NO production [13]. NO may also act at the presynaptic terminal, and induction of LTP in single synaptic pairs in dissociated hippocampal cultures can be blocked by the presence of an NO-sequestering agent in the presynaptic cell [15]. Taken together, these data all indicate that NO is synthesized in the postsynaptic neuron following tetanic stimulation or paired depolarization, and participates in the induction of LTP which can be expressed both in neighbouring neurons and in presynaptic terminals. Moreover, although NO itself cannot induce synaptic potentiation, it can convert weak, non-LTP-inducing tetanic stimulation into long-lasting potentiation [13], indicating that NO may need to combine with some other signal, such as a rise in intracellular calcium, in the postsynaptic or presynaptic cell in order to generate LTP.

More recently, transgenic 'knock-out' mice have been generated that lack the nNOS or eNOS gene, or both. These studies have proved to be very revealing, as LTP induction in slices of hippocampus from nNOS knock-out mice appeared to be completely normal [10]. Furthermore, LTP induction in these mice could be blocked by NOS inhibitors. This was rather a surprise, as it was thought that nNOS was the only NOS isoform present in neurons. Subsequent staining with an anti-eNOS antibody, however, revealed high levels of eNOS in hippocampal pyramidal neurons from both the knock-out and wild-type mice [10]. This stunning result suggests that eNOS is present in pyramidal neurons and may be the isoform responsible, alone or in concert with nNOS, for NO production during LTP induction. Transgenic mice were then created that lacked both eNOS and nNOS (double knock-out), and LTP was significantly reduced in hippocampal slices from these animals [11], providing strong evidence that LTP induction in hippocampal CA1 neurons requires the production of NO. Moreover, it is likely that the eNOS isoform may be dominant in LTP induction in normal hippocampal slices. Overexpression (via an adenovirus expression vector) of a myristoylation-deficient form of eNOS prevents the induction of LTP in CA1 neurons [6]. LTP can be induced, however, if the mutated eNOS is tagged with CD8 (a cell surface protein which inserts into the plasma membrane, dragging the eNOS along as well). Furthermore, LTP under these conditions can be blocked by NOS inhibitors. These data suggest that the eNOS isoform is sufficient to induce LTP and that the enzyme needs to be associated with the plasma membrane in order to function (Figure 4).

It is now becoming clear that NO production is not required for all forms of LTP in CA1 neurons and that NO-dependent and -independent forms of LTP exist. In the apical dendrites of the stratum radiatum (Figure 2), NO-independent LTP can be induced by altering the experimental recording conditions [13]. Furthermore, in the basal dendrites of the stratum oriens only NO-independent LTP is observed, despite the presynaptic neurons arising from the same source in the stratum oriens and the stratum radiatum [16]. This astonishing finding appears to arise from a differential distribution of eNOS in the postsynaptic CA1 pyramidal neuron; it is present in apical but not basal dendrites [10]. The implications arising from the difference in apical and basal LTP mechanisms are interesting, as the majority of the connections on to the stratum radiatum CA1 neuron arise from the ipsilateral hippocampus, while the connections in the stratum oriens arise mainly from the contralateral hippocampus. Is it possible, therefore, that these different LTP induction mechanisms could serve to distinguish between CA3 inputs from different sides of the brain?

Following the discovery that NO may be the retrograde messenger in LTP, interest has also focused on another gaseous molecule, CO. The evidence in support of CO involvement in LTP induction is, however, weak when compared with that for NO. The inhibitors of HO are not selective (see above) and, although a block of LTP induction is seen with some HO inhibitors [17], this does not correlate with their ability to decrease CO production from hippocampal slices [18]. Furthermore, LTP in knock-out mice lacking the *HO-2* gene is normal and, although LTP induction in these mice can be prevented by a HO inhibitor [19], it is unlikely that compensation by other HO isoforms is occurring, as there is only one constitutive form of HO. It is more probable, therefore, that the inhibitor is acting on a non-HO target, possibly GC. On the available evidence it therefore seems unlikely that CO is a messenger in hippocampal LTP.

The mechanism by which NO contributes to the stabilization of LTP during induction is not really known. One target of NO is GC and, following tetanic stimulation, cGMP levels in hippocampal slices are increased.

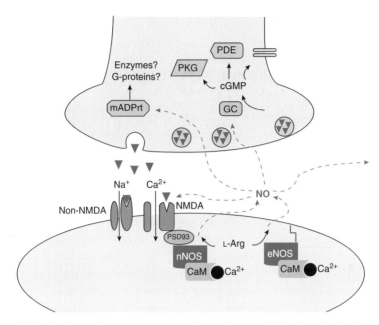

Figure 4. Mechanisms that may be occurring during and following the induction of LTP at the CA3 to CA1 synapse

Following calcium influx into the presynaptic terminal, glutamate (dark blue triangles) is released from the nerve terminals. Under normal conditions this activates non-NMDA receptors, but, following tetanic stimulation or depolarization of the postsynaptic cell, the NMDA receptor is also activated and calcium flows into the postsynaptic cell. This calcium binds to calmodulin (CaM), which in turn binds to and activates nNOS and/or eNOS. Both NOS isoforms may be held close to the plasma membrane, eNOS via its myristoylation site (zigzag line) and nNOS perhaps by interaction with a postsynaptic density protein (PSD93). NO is liberated from L-arginine (L-Arg), and this diffuses freely to targets in the presynaptic terminal or neighbouring postsynaptic neurons (not shown). These targets may include GC or mono-ADP-ribosyltransferase (mADPrt). GC activation results in cGMP production, and this in turn could activate cGMP-dependent protein kinase (PKG) or a cGMP-stimulated cyclic nucleotide phosphodiesterase (PDE), or open ion channels. The result is an increase in the response of the postsynaptic neuron to synaptic stimulation; the final mechanism by which this occurs is still unknown.

Membrane-permeable cGMP analogues can rescue LTP that has been inhibited by NOS inhibitors [13], and in dissociated hippocampal cultures injection of cGMP into the presynaptic neuron results in LTP when paired with a weak non-LTP-inducing tetanus [20]. This suggests that cGMP production in the presynaptic terminal may be the means by which NO induces LTP (Figure 4). As with NO, some synaptic activity in the neuron is necessary to obtain long-lasting potentiation, as neither NO nor cGMP can elicit potentiation on their own. Another possible mechanism is that of NO-induced auto-ADP-ribosylation of glyceraldehyde-3-phosphate dehydrogenase (Figure 4). Following tetanic stimulation there is an increase in ADP-ribosylation in hippocampal slices, and inhibitors of mono-ADP-ribosyltransferase can also prevent the induction of LTP [13]. Regardless of whether the effects of NO occur through

cGMP production or ADP-ribosylation, the final target (e.g. release machinery or receptor) remains unknown.

Do NO and CO contribute to nociceptive signalling within the spinal cord?

Hyperalgesia and sensitization within the spinal cord also involve the activation of NMDA receptors by glutamatergic sensory fibres and result in
increased excitability of dorsal horn neurons. This can occur following peripheral inflammation produced by chemical agents such as formalin or
carrageenan or by peripheral nerve damage, and can be mimicked by the
intrathecal administration of NMDA. This form of plasticity within the spinal
cord could occur during some chronic pain states, and the underlying mechanisms involved are therefore of considerable interest to clinicians and scientists. Given that both the induction of LTP and spinal sensitization appear to
require NMDA receptor activation, it is possible that they have other mechanisms in common, perhaps including NO generation. NO is also unlikely to
be a 'neurotransmitter' in the spinal cord, as intrathecal administration of NOS
inhibitors does not alter acute nociceptive responses to heat or to mechanical
or electrical stimulation [21]. NO may, however, be involved in the facilitation
of responses caused by peripheral inflammation or nerve damage; following
ligation of the sciatic nerve or the intrathecal administration of NMDA, mice
appear hyperalgesic and withdraw their tail or paw much faster in response to
applied heat or pressure. Intrathecal administration of NOS inhibitors prevents this sensitization from occurring, and the paw/tail withdrawal latencies
are the same as for controls [21]. Intrathecal application of NO donor compounds also results in hyperalgesia, mimicking intrathecal NMDA administration. Spontaneous licking of the paw in response to a formalin injection (a
response that results from the development of inflammation in the paw and
activation of spinal NMDA receptors) is also prevented by NOS inhibitors
[21]. These responses in behavioural studies are mirrored in electrophysiological studies, where the increase in dorsal horn neuron firing following injection
of formalin into the paw can be prevented by the intrathecal application of
NOS inhibitors. These findings each imply that NO within the spinal cord
participates in the hyperalgesia generated by peripheral insults or the spinal
application of NMDA. Furthermore, NO may diffuse to act on neighbouring
cells and terminals, as intrathecal haemoglobin prevents the development of
hyperalgesia following spinal NMDA administration [21]. NO may act on GC
to elicit hyperalgesia, as GC inhibitors prevent the generation of hyperalgesia
following peripheral nerve ligation or intrathecal NMDA administration [21],
although it is not known whether cGMP alone can elicit hyperalgesia.

So, all the evidence from studies using NOS and GC inhibitors suggests
that NO is produced, following NMDA receptor activation, as a result of
peripheral nerve damage, chemical insult or direct application of intrathecal

NMDA agonists, and that NO may activate GC, resulting in hyperalgesia. Unfortunately, a study with nNOS knock-out mice has found that formalin-induced licking behaviour is not altered when nNOS is lacking [22]. Furthermore, NOS inhibitors are without effect in these mice, eliminating up-regulation of eNOS as an explanation. It appears that some other compensation mechanism may be occurring in these mice; perhaps a molecule such as CO may be fulfiling the role vacated by NO. Whether CO has any role in spinal nociception in non-transgenic animals is unclear, as all studies to date have used metalloporphyrin inhibitors of HO and, as mentioned above, these are non-specific, inhibiting both NOS and GC. These HO inhibitors do not mimic NOS inhibitors completely, however, as they prevent the generation of hyperalgesia to mechanical stimuli in preference to thermal stimuli, whereas NOS inhibitors are very effective at preventing thermal hyperalgesia [23]. This suggests that CO may possibly contribute to some forms of hyperalgesia. Further work is therefore required to determine whether NO or CO (or both) contribute to the development of hyperalgesia within the spinal cord.

It is clear that, within the CNS, NO is probably not a neurotransmitter in the classical sense, as it is neither stored in nor released from vesicles in nerve terminals, and it does not participate in synaptic transmission. Nonetheless, it remains an important modulator at some synapses in the CNS and may contribute to various forms of synaptic plasticity that could underlie learning, memory and chronic pain. The role of CO within the CNS is less well defined, and this mainly results from the lack of selective HO-2 inhibitors available; hopefully the development of better tools will allow us to probe the role of CO in the CNS further. Finally, it should not be forgotten that there may well be other members of this family of gaseous molecules, yet to be discovered, that may also contribute to signalling within the CNS.

Summary

- *NO and CO are small gaseous molecules that can be synthesized de novo in neuronal tissue and can diffuse readily through the plasma membrane.*
- *NOS inhibitors prevent the induction of LTP in the hippocampus, and studies with NOS knock-out mice and viral overexpression of mutated NOS indicate that the endothelial form of the enzyme is probably responsible for NO production in these neurons.*
- *Inhibitors of CO production can block the induction of LTP, but this does not correlate with their ability to prevent CO production in the hippocampus. LTP is normal in mice that lack HO-2 and, furthermore, there is no obvious mechanism by which HO could be activated during synaptic stimulation.*

- *NO probably diffuses out of the postsynaptic neuron and acts on neighbouring neurons and presynaptic terminals to either instigate, or assist in, the generation or stabilization of LTP, possibly by activating GC.*
- *There are NO-dependent and NO-independent forms of LTP, and both forms can be found at synapses on to the same neuron. It is therefore possible that subtle discrimination can occur between different inputs on to the same cell.*
- *NO may also participate in the induction of sensitization within the spinal cord. NOS inhibitors can prevent the development of spinal hyperalgesia due to intrathecal NMDA administration or peripheral nerve injury, and could therefore contribute to some chronic pain states.*

References

1. Furchgott, R.F. & Zawadski, J.V. (1980) The obligatory role of endothelial cells in the relaxation of arterial smooth muscle by acetylcholine. *Nature (London)* **288**, 373–376
2. Palmer, R.M.J., Ferrige, A.G. & Moncada, S. (1987) Nitric oxide release accounts for the biological activity of endothelium-derived relaxing factor. *Nature (London)* **327**, 524–526
3. Garthwaite, J. & Boulton, C.L. (1995) Nitric oxide signaling in the central nervous system. *Annu. Rev. Physiol.* **57**, 683–706
4. Rand, M.J. & Li, C.G. (1995) Nitric oxide as a neurotransmitter in peripheral nerves: nature of transmitter and mechanism of transmission. *Annu. Rev. Physiol.* **57**, 659–682
5. Förstermann, U. & Kleinert, H. (1995) Nitric oxide synthase: expression and expressional control of the three isoforms. *Naunyn-Schmiedeberg's Arch. Pharmacol.* **325**, 351–364
6. Kantor, D.B., Lanzrein, M., Stary, J., Sandoval, G.M., Smith, W.B., Sullivan, B.M., Davidson, N. & Schuman, E.M. (1996) A role for endothelial NO synthase in LTP revealed by adenovirus-mediated inhibition and rescue. *Science* **274**, 1744–1748
7. Brenman, J.E., Christopherson, K.S., Craven, S.E., McGee, A.W. & Bredt, D.S. (1996) Cloning and characterization of postsynaptic density 93, a nitric oxide synthase interacting protein. *J. Neurosci.* **16**, 7407–7415
8. Moore, P.K. & Handy, R.L.C. (1997) Selective inhibitors of neuronal nitric oxide synthase – is no NOS really good NOS for the nervous system. *Trends Pharmacol. Sci.* **18**, 204–211
9. Maines, M.D. (1997) The heme oxygenase system: A regulator of second messenger gases. *Annu. Rev. Pharmacol. Toxicol.* **37**, 517–554
10. O'Dell, T.J., Huang, P.L., Dawson, T.M., Dinnerman, J.L., Snyder, S.H., Kandel, E.R. & Fishman, M.C. (1994) Endothelial NOS and the blockade of LTP by NOS inhibitors in mice lacking neuronal NOS. *Science* **265**, 542–546
11. Son, H., Hawkins, R.D., Martin, K., Kiebler, M., Huang, P.L., Fishman, M.C. & Kandel, E.R. (1996) Long-term potentiation is reduced in mice that are doubly mutant in endothelial and neuronal nitric oxide synthase. *Cell* **87**, 1015–1023
12. Dun, N.J., Dun, S.L., Wu, S.Y., Förstermann, U., Schmidt, H.H.H.W. & Tseng, L.F. (1993) Nitric oxide synthase immunoreactivity in the rat, mouse, cat and squirrel monkey spinal cord. *Neuroscience* **53**, 845–857
13. Schuman, E.M. & Madison, D.V. (1994) Nitric oxide and synaptic function. *Annu. Rev. Neurosci.* **17**, 153–183
14. Lledo, P.-M., Zhang, X., Südhof, T.C., Malenka, R.C. & Nicoll, R.A. (1998) Postsynaptic membrane fusion and long-term potentiation. *Science* **279**, 399–403

15. Arancio, O., Kiebler, M., Lee, C.J., Lev-Ram, V., Tsien, R.Y., Kandel, E.R. & Hawkins, R.D. (1996) Nitric oxide acts directly in the presynaptic neuron to produce long-term potentiation in cultured hippocampal neurons. *Cell* **87**, 1025–1035

16. Haley, J.E., Schiable, E., Pavlidis, P., Murdock, A. & Madison, D.V. (1996) Basal and apical synapses of CA1 pyramidal cells employ different LTP induction mechanisms. *Learn. Mem.* **3**, 289–295

17. Zhuo, M., Small, S.A., Kandel, E.R. & Hawkins, R.D. (1993) Nitric oxide and carbon monoxide produce activity-dependent long-term synaptic enhancement in hippocampus. *Science* **260**, 1946–1950

18. Meffert, M.K., Haley, J.E., Schuman, E.M., Schulman, H. & Madison, D.V. (1994) Inhibition of hippocampal heme oxygenase, nitric oxide synthase, and long-term potentiation by metalloporphyrins. *Neuron* **13**, 1225–1233

19. Poss, K.D., Thomas, M.J., Ebralidze, A.K., O'Dell, T.J. & Tonegawa, S. (1995) Hippocampal long-term potentiation is normal in heme oxygenase-2 mutant mice. *Neuron* **15**, 867–873

20. Arancio, O., Kandel, E.R. & Hawkins, R.D. (1995) Activity-dependent long-term enhancement of transmitter release by presynaptic 3',5'-cyclic GMP in cultured hippocampal neurons. *Nature (London)* **376**, 74–80

21. Meller, S.T. & Gebhart, G.F. (1993) Nitric oxide (NO) and nociceptive processing in the spinal cord. *Pain* **52**, 127–136

22. Crosby, G., Marota, J.J.A. & Huang, P.L. (1995) Intact nociception-induced neuroplasticity in transgenic mice deficient in neuronal nitric oxide synthase. *Neuroscience* **69**, 1013–1017

23. Meller, S.T., Dykstra, C.L. & Gebhart, G.F. (1994) Investigations of the possible role for carbon monoxide (CO) in thermal and mechanical hyperalgesia in the rat. *NeuroReport* **5**, 2337–2341

8

Molecular biology of olfactory receptors

Yitzhak Pilpel, Alona Sosinsky & Doron Lancet[1]

Department of Molecular Genetics, The Weizmann Institute of Science, Rehovot 76100, Israel

Overview

In order to elicit an olfactory response, a substance has to partition into the gas phase and diffuse into the nose. Such odorant molecules, usually low-molecular-mass hydrophobic compounds, encounter the ciliated endings of sensory neuronal dendrites, which protrude into a mucus layer at the surface of the olfactory epithelium in the nasal cavity. Embedded in the membranes of such cilia are olfactory receptor (OR) proteins, which recognize odorants and elicit a transduction cascade that underlies the nerve cell response. The sensory axons project to the olfactory bulb in the brain, where they converge into synaptic structures called glomeruli. The specific convergence patterns of olfactory axons, which depend on OR expression, provide a model system for neuronal network development. Here, initial processing of odour information occurs, which is followed by additional analysis in higher olfactory brain centres.

Chemical detection in a probabilistic receptor repertoire

Most biological recognition systems have evolved towards an optimized specificity for endogenous ligands, such as hormones, neurotransmitters and enzyme substrates. Perhaps the most peculiar aspect of the olfactory system is that it has to bind, and uniquely recognize, a vast array of ligands, most of which are xenobiotic in origin. By analogy with the immune system, it was proposed that the olfactory pathway must have evolved a large repertoire of potential ORs,

[1]*To whom correspondence should be addressed.*

---

most of which may have arisen without *a priori* specificity towards particular odorants [1]. A probabilistic receptor affinity distribution model has been developed, which relates the number of different receptor types required to ensure binding of any odorant with a minimal physiologically significant affinity [2]. This allowed an early prediction that the olfactory repertoire in mammalian species contains approx. 1000 different proteins [1], in good agreement with later experimental estimates [3].

Prior to the discovery of the first ORs, it had been established by electrophysiological recordings that each olfactory sensory neuron may respond to a wide range of odorant chemicals [4]. This raised two alternative explanations: that each neuron expresses a number of different, narrowly tuned ORs; or that a neuron expresses only one OR type, but with an unusually broad specificity. It is currently believed that the latter situation prevails, i.e. that OR expression is 'clonally excluded' (see below). With the number of possible odorants far exceeding that of OR types, it is obvious that each sensory neuron contributes to the recognition of numerous chemical substances. However, each odorant is unequivocally identified through its eliciting a unique 'across-neuron' pattern, as demonstrated clearly when numerous single-unit recordings are analysed [4].

ORs belong to the G-protein-coupled receptor (GPCR) hyperfamily

The earlier demonstration that odorant responses are mediated by GTP-binding proteins and adenylate cyclase activation raised the hypothesis that ORs might belong to the GPCR hyperfamily of proteins [5], all of which share seven hydrophobic transmembrane helices. Such receptors have been implicated in numerous cellular transduction processes, including photoreception, as well as hormone and neurotransmitter binding [6]. Accordingly, the first cloning of candidate OR genes from rat olfactory epithelium was accomplished through the use of degenerate GPCR-conserved sequences as PCR primers [3]. This was then followed by the cloning of such genes in several species, including human [7,8], mouse [9], dog [7] and fish [10]. OR genes from all vertebrate species constitute a superfamily whose members share at least 25% amino acid sequence identity [11].

Analysis of numerous OR sequences from a variety of species clearly confirms their identity as GPCRs (Figure 1a). On the other hand, several features appear to be unique to the OR protein sequences. The most prominent hallmarks are the conserved regions in the second and third intracellular loops (Y. Pilpel and D. Lancet, unpublished work; Figure 1a). These may be implicated as molecular interfaces with the olfactory G-protein.

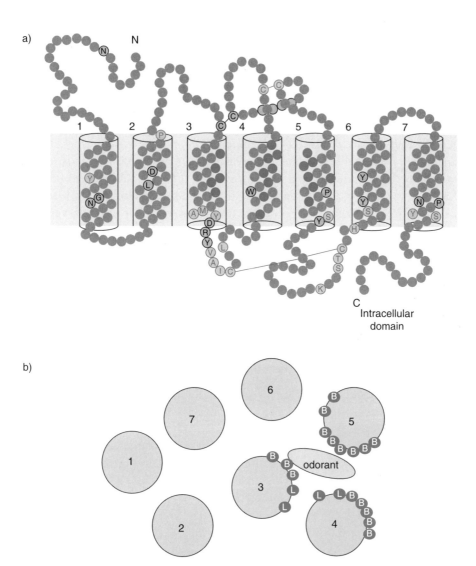

Figure 1. (a) Schematic two-dimensional diagram of the OR protein, and (b) cross-sectional representation of the OR protein in the cell membrane (extracellular view) (a) The 20 variable residues that constitute the putative complementarity-determining region on transmembrane helices 3, 4, and 5 are shown as solid blue circles. These residues are mainly clustered along one side of each of the three variable helices. Highly conserved residues, marked with the single-letter amino acid code, are either conserved in all GPCRs (circled in blue) or unique to ORs (circled in black). These include the conserved glycosylation site at an asparagine residue near the N-terminus (marked by N) and the DRY motif at the end of the third transmembrane helix. Putative disulphide bridges are indicated with a black line (for those unique to ORs [12]) or a blue line (conserved in all GPCRs). (b) Looking down on the transmembrane helices (1–7), the residues comprising the putative complementarity-determining region are shown as dark blue circles. Residues with hydrophobic and hydrophilic side chains are marked B and L respectively. A schematic odorant is shown in the putative binding site.

Odorant complementarity-determining regions

A wealth of pharmacological and site-directed mutagenesis experiments on several GPCRs established that the binding of low-molecular-mass ligands takes place in the plane of the membrane, within the barrel enclosed by the seven helices [6]. While no analogous experimental information is yet available for OR proteins, a computational approach has been developed for these proteins, based on the availability of numerous homologous sequences. Initial analyses indicated that transmembrane helices 3, 4 and 5 harbour most of the variability [3] (Figure 1a), and it was suggested that they take part in odorant binding, subserving a function similar to the immunoglobulin hypervariable domains [5]. However, it remained unknown whether the OR variable residues are clustered in space. In a recent study [12], analysis of hundreds of OR sequences, along with molecular modelling of the receptor structure, revealed a unique pattern of diversity, whereby most of the variable amino acid residues are clustered in a specific region in the interior of the seven-helix barrel (Figure 1b). The set of 20 interior-facing variable residues thus defined was proposed to serve as the complementarity-determining region for odour recognition (Figure 1b).

Evolution of the OR repertoire

In mammalian species, the OR gene superfamily is estimated to consist of several hundred genes [3], with a relatively large percentage of pseudogenes, especially in the human [13]. This gene repertoire, which constitutes the 'olfactory sub-genome' [11], is estimated to encompass ~1% of the entire genome of mammalian species. The open reading frames are found in several chromosomes, arranged in clusters of 10 or more members [8]. Unlike the somatic gene recombination and mutation mechanisms that account for immune diversity, OR repertoire diversity seems to be germ-line inherited, since the OR coding regions are uninterrupted in the genome [3,14]. The OR repertoire, similar to that of other multigene families, is likely to have arisen by a long evolutionary process of gene duplication, followed by germ-line mutations. This process of repertoire expansion is likely to have been driven by the need to enhance the chances of a successful probabilistic odorant binding.

The biochemical cascade in olfactory signalling

While odorant binding to an OR may be considered as an input 'written' in a chemical language, a corresponding output must be generated which translates the information into electrophysiological signals. The relevant signal transduction cascade has been shown to include a stimulatory GTP-binding protein, adenylate cyclase and a cAMP-gated cation channel [15] (Figure 2). Olfactory transduction manifests an aspect of universality by involving the same primary components as many other neural systems with seven-helix receptors, includ-

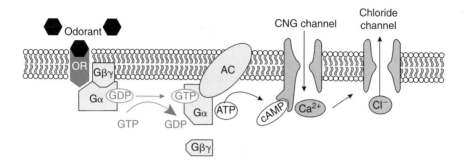

Figure 2. Odorant-dependent signal transduction pathway
Following the binding of an odorant to its receptor (OR), the olfactory G-protein exchanges GDP for GTP. Concomitantly the α subunit (G_α) dissociates from the βγ subunits ($G_{\beta\gamma}$) and activates adenylate cyclase type III (AC), which generates cAMP from ATP. cAMP opens the cyclic-nucleotide-gated (CNG) cation channel for passage of Ca^{2+} ions, resulting in cell depolarization and the generation of an action potential. The current is amplified by activation of a Ca^{2+}-dependent chloride channel.

ing neurotransmitter and visual receptors [16]. On the other hand, it seems to share only with the visual system the involvement of a direct gating of the ion channel by the cyclic nucleotide second messenger [17]. While the overall features of the signal transduction pathway are identical to those of other signalling systems, the specific components are olfactory-unique and highly tissue- and cilia-enriched. The cloned olfactory specific transduction proteins are G_{olf} (a G_s-like protein), adenylate cyclase III and the cAMP-gated channel.

The central role of cAMP and G-proteins in the olfactory pathway was substantiated by several independent approaches. A strong positive correlation exists between the magnitude of adenylate cyclase stimulation and the summated electrophysiological responses [18]. In addition, both cAMP derivatives and guanine nucleotides modulate or mimic odorant responses in epithelial sensory neurons [19] and in cultured olfactory cells [20]. Perhaps the most clear-cut evidence for the primary significance of the cAMP pathway to mammalian olfactory reception has been obtained recently with knock-out mice lacking either the functional cyclic-nucleotide channel [21] or the olfactory G-protein (G_{olf}) [22]. Both mutant strains exhibited a dramatic decrease in the primary electrophysiological response to all odorants tested, and revealed behavioural phenotypes consistent with general anosmia.

The multi-step olfactory transduction cascade has been implicated in signal amplification. Odorant receptors are rather broadly specific, and may typically have rather low ligand affinities. However, when coupled to a system where a single activated receptor may catalyse the generation of thousands of second-messenger molecules, an ultra-high sensitivity may materialize [1,5]. In this, an analogy may be drawn with other sensory pathways, including vision and some taste pathways [23].

An alternative olfactory transduction pathway has been proposed, involving the second messenger inositol 1,4,5-trisphosphate ($InsP_3$). The evidence was based on rapid kinetic measurements in isolated olfactory cilia [24] and OR-expressing cells [25], showing $InsP_3$-mediated responses to certain classes of odorant. However, the 'odorant pharmacology' experiments [18], and most convincingly the gene inactivation experiments [21,22], demonstrated that cAMP is very likely to also mediate the responses to the $InsP_3$-related odorants. It may therefore be concluded that cAMP is a necessary second messenger for all or most odorants in mammals.

Expressed OR proteins and their ligand specificity

For years OR genes constituted 'semi-orphan' receptors: as a group, they were assigned a very likely function in odorant recognition, yet a detailed assignment of odorant specificity to individual OR genes remained largely lacking. This has begun to change rapidly, starting with the first identification of a functional odorant receptor, coded by the *odr-10* gene, in the nematode *Caenorhabditis elegans*, identified in a behavioural mutant incapable of detecting the odorant diacetyl [26]. Odr-10 functionality was further corroborated in a human expression system, where the receptor was shown to activate Ca^{2+} release in response to diacetyl [27].

Mammalian OR gene expression was first described in a baculovirus system [25,28], including the identification and purification of the encoded 30 kDa OR polypeptide. Following the earlier observation of odorant responses in such OR-expressing insect cells [25], the expressed protein was reconstituted in lipid vesicles and binding of similar odorants was observed by photoaffinity labelling [29]. Subsequently, three zebrafish odorant receptors were expressed in a human cell line, resulting in a transient increase in intracellular Ca^{2+} in response to odorants in fish-food extracts [30].

More recently, Zhao et al. [31] used rat nasal epithelium as an *in vivo* OR expression system, employing recombinant adenovirus infection. Electrophysiological recordings showed that, due to the expression of the rat receptor *I7* gene by an increased number of olfactory neurons, a 4–7-fold enhancement was effected in the responses to the odorant octyl aldehyde (octanal), as well as to several structurally related compounds. A carefully documented dose–response curve lends credence to the notion that OR gene expression has been achieved.

Another strategy employed for the identification of functional ORs was mutation analysis and genetic linkage. In a study of two inbred mouse strains differing in their sensitivity to isovaleric acid, genetic linkage was established between odorant sensitivity and markers on chromosomes 4 and 6. It was concluded that the most likely cause of the inability to sense isovaleric acid is the loss of receptor protein(s) residing in the vicinity of these markers. A similar linkage analysis has been conducted in humans, resulting in a tentative assignment of odorant sensitivity traits to distinct OR clusters in the human genome (S. Horn-Saban and D. Lancet, unpublished work).

Patterns of olfactory receptor expression and their transcriptional regulation

The patterns of expression of different odorant receptor genes within the olfactory epithelium of rodents have been examined in a series of *in situ* hybridization experiments [32,33]. A given OR probe was found to identify about 0.1% of the sensory neurons. Given an estimate of 500–1000 OR genes in the rat genome [3], these findings are consistent with the hypothesis of OR clonal exclusion, i.e. each neuron expresses a single receptor gene [34]. Sensory neurons expressing the same receptor or receptor subset appear to be topologically segregated into one of four broad zones extending along the anterior–posterior axis of the nasal cavity (Figure 3). These zones exhibit bilateral symmetry in the two nasal cavities. Within a given zone, however, olfactory neurons expressing a given receptor appear to be distributed randomly rather than spatially localized. Moreover, RNA from cells heterozygous for an OR gene hybridizes with only one of the allele probes [35], suggesting that only one OR allele can be activated in a given neuron. In summary, since each neuron may express only one OR protein from a single allele, the task of odour-quality perception may be reduced to detecting which subset of sensory neurons has been activated [1,33,34].

The functional significance of clonal exclusion was demonstrated recently in an elegant analysis of olfactory-dependent chemotaxis in *C. elegans* [36]. It was shown that nematodes that transgenically expressed the chemoattractive diacetyl receptor Odr-10 in neurons that normally mediate odorant-induced repulsion were repelled by diacetyl. It was thus concluded that the cellular context of the activated receptor, and not the receptor itself, determines the final odour-quality perception and the subsequent behavioural response.

What is the mechanism that controls this complex pattern of odorant receptor expression? A recent study, using a transgenic animal approach, has shown that a 6.7 kb region upstream of the mouse M4 OR coding region is sufficient to direct several aspects of OR-regulated expression [37]. This includes the specificity to olfactory epithelial tissue, the restricted expression in only one of the epithelial zones, and excluded expression in a small randomly disposed subset of the cells within a zone. This is in line with reports of a 4–6 kb 5' intron that is immediately preceded by a non-coding exon and a putative upstream gene-control region [14,37]. However, the transgenic 6.7 kb upstream region was not sufficient to dictate expression in a specific zone in which a given OR is known to be expressed naturally, and did not include the potential feedback loop that prevents individual sensory neurons from expressing more than one OR gene. Such specific orchestration is likely to arise from a more complex, hierarchical series of regulated transcriptional controls.

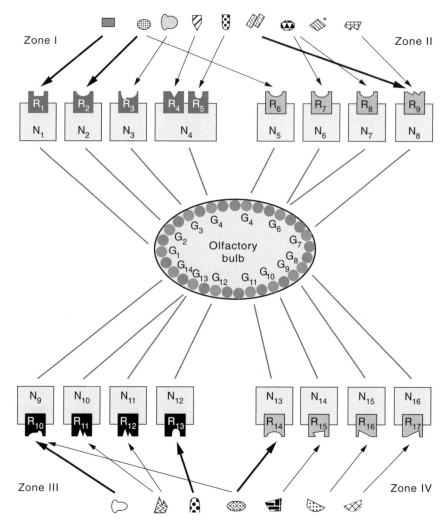

Figure 3. Schematic view of neuronal odour recognition and coding apparatus
OR-expressing neurons are designated by N_i. Each neuron may express one type of OR (in multiple copies), or perhaps very few (e.g. N_4). However, the same OR is expressed by several neurons, scattered randomly in one of four zones of the olfactory epithelium (all ORs expressed at a given zone are shown in the same colour). An array of odour molecules may bind the ORs with different affinities (shown by the size of the arrows) (e.g. R_7 and R_8); conversely, the same odour molecule can bind to different receptors with different affinities (R_2 and R_5). An odorant with no OR binder (*) may not be detected by this system. Neurons expressing identical ORs (N_6 and N_7) project axons to the same glomerulus (G_6) of the olfactory bulb.

An attractive model [32,35,37] is that, during development, spatially restricted transcription factors activate a subpopulation of genes within an active chromosomal OR cluster, so as to define the repertoire of receptor genes that can be expressed within a topological zone. These factors could bind to distant regions (e.g. enhancers), functioning perhaps in co-operation with the elements located within the proximal promoter. Subsequent selection of the receptor(s) to be expressed in a single olfactory neuron could result from competition for *trans*-acting factors or relief from negative regulation mediated by DNA methylation or inhibitory factors. Finally, a feedback mechanism could serve to restrict expression to a single functional receptor in each neuron [3].

Olfactory bulb glomeruli represent ORs

The OR-expressing neurons project their axons to the olfactory bulb, where they converge on to synaptic complexes termed glomeruli (Figure 3). Each of the roughly 1000 glomeruli receives input from about 10000 sensory neurons. Early analyses of this convergence pattern led to the idea that glomeruli serve as 'functional addresses' [38] and, more specifically, that each glomerulus subserves all the neurons that express a particular OR [34]. This was later confirmed experimentally by showing that all olfactory neurons in the mouse expressing a given OR gene project axons to only two topographically fixed glomeruli in each of the two olfactory bulbs [39]. Further experiments, in which one OR coding region was swapped with another, led to axonal convergence on to a glomerular target different from that of both receptors. It was then concluded that the OR coding region itself carries necessary, but probably not sufficient, information required for axonal guidance. Furthermore, sequence analysis of several ORs identified a specific region of the OR molecule, on the second extracellular loop, that may be responsible for such axonal guidance [40].

What is the logic behind this situation? It appears that an olfactory neuron has to make two major decisions during its lifetime, namely which OR to express and into which glomerulus to project its axon. If the choices are co-ordinated, a link is formed between ligand specificity and neuronal processing. Observations obtained from 2-deoxyglucose measurements [38] and single-unit recording of neurons from the olfactory bulb [41] support this notion. It was shown that glomerular cells at given bulb loci respond to odorants with well-defined chemical sub-modalities, size and structure, potentially recognized by the same, or similar, ORs. The relationship between OR function and central nervous system addressing allows thousands of similar neurons to converge upon a restricted subset of secondary neuronal targets. This allows signal averaging and noise reduction, as well as the formation of a functionally consistent map in the brain [38].

Summary

- *OR proteins bind odorant ligands and transmit a G-protein-mediated intracellular signal, resulting in generation of an action potential.*
- *The accumulation of DNA sequences of hundreds of OR genes provides an opportunity to predict features related to their structure, function and evolutionary diversification.*
- *The OR repertoire has evolved a variable ligand-binding site that ascertains recognition of multiple odorants, coupled to constant regions that mediate the cAMP-mediated signal transduction.*
- *The cellular second messenger underlies the responses to diverse odorants through the direct gating of olfactory-specific cation channels.*
- *This situation necessitates a mechanism of cellular exclusion, whereby each sensory neuron expresses only one receptor type, which in turn influences axonal projections.*
- *A 'synaptic image' of the OR repertoire thus encodes the detected odorant in the central nervous system.*

References

1. Lancet, D. (1986) Vertebrate olfactory reception. *Annu. Rev. Neurosci.* **9**, 329–355
2. Lancet, D., Sadovsky, E. & Seidmann, E. (1993) Probability model for molecular recognition in biological receptor repertoires: significance to the olfactory system. *Proc. Natl. Acad. Sci. U.S.A.* **90**, 3715–3719
3. Buck, L. & Axel, R. (1991) A novel multigene family may encode odorant receptors: a molecular basis for odor recognition. *Cell* **65**, 175–187
4. Sicard, G. & Holley, A. (1984) Receptor cell responses to odorants: similarities and differences among odorants. *Brain Res.* **292**, 283–296
5. Lancet, D. & Pace, U. (1987) The molecular basis of odor recognition. *Trends Biochem. Sci.* **12**, 63–66
6. Baldwin, J.M. (1994) Structure and function of receptors coupled to G-proteins. *Curr. Opin. Cell Biol.* **6**, 180–190
7. Parmentier, M., Libert, F., Schurmans, S., Schiffman, S., Lefort, A., Eggerickx, D., Ledent, C., Mollereau, C., Gerard, C., Perret, J., et al. (1992) Expression of members of the putative olfactory receptor gene family in mammalian germ cells. *Nature (London)* **355**, 453–455
8. Ben-Arie, N., Lancet, D., Taylor, C., Khen, M., Walker, N., Ledbetter, D.H., Carrozzo, R., Patel, K., Sheer, D., Lehrach, H., et al. (1994) Olfactory receptor gene cluster on human chromosome 17: possible duplication of an ancestral receptor repertoire. *Hum. Mol. Genet.* **3**, 229–235
9. Sullivan, S.L., Adamson, M.C., Ressler, K.J., Kozak, C.A. & Buck, L.B. (1996) The chromosomal distribution of mouse odorant receptor genes. *Proc. Natl. Acad. Sci. U.S.A.* **93**, 884–888
10. Ngai, J., Dowling, M.M., Buck, L., Axel, R. & Chess, A. (1993) The family of genes encoding odorant receptors in the channel catfish. *Cell* **72**, 657–666
11. Lancet, D. & Ben-Arie, N. (1993) Olfactory receptors. *Curr. Biol.* **3**, 668–674
12. Reference deleted
13. Rouquier, S., Taviaux, S., Trask, B.J., Brand-Arpon, V., van den Engh, G., Demaille, J. & Giorgi, D. (1998) Distribution of olfactory receptor genes in the human genome. *Nature Genet.* **18**, 243–250

14. Glusman, G., Clifton, S., Roe, R. & Lancet, D. (1996) Sequence analysis in the olfactory receptor gene cluster on human chromosome 17: recombinatorial events affecting receptor diversity. *Genomics* **37**, 147–160

15. Reed, R.R. (1994) The molecular basis of sensitivity and specificity in olfaction. *Semin. Cell Biol.* **5**, 33–38

16. Dohlman, H.G., Thorner, J., Caron, M.G. & Lefkowitz, R.J. (1991) Model systems for the study of seven-transmembrane-segment receptors. *Annu. Rev. Biochem.* **60**, 653–688

17. Torre, V., Ashmore, J.F., Lamb, T.D. & Menini, A. (1995) Transduction and adaptation in sensory receptor cells. *J. Neurosci.* **15**, 7757–7768

18. Lowe, G., Nakamura, T. & Gold, G.H. (1989) Adenylate cyclase mediates olfactory transduction for a wide variety of odorants. *Proc. Natl. Acad. Sci. U.S.A.* **86**, 5641–5645

19. Firestein, S., Darrow, B. & Shepherd, G.M. (1991) Activation of the sensory current in salamander olfactory receptor neurons depends on a G protein-mediated cAMP second messenger system. *Neuron* **6**, 825–835

20. Ronnett, G.V., Parfitt, D.J., Hester, L.D. & Snyder, S.H. (1991) Odorant-sensitive adenylate cyclase: rapid, potent activation and desensitization in primary olfactory neuronal cultures. *Proc. Natl. Acad. Sci. U.S.A.* **88**, 2366–2369

21. Brunet, L.J., Gold, G.H. & Ngai, J. (1996) General anosmia caused by a targeted disruption of the mouse olfactory cyclic nucleotide-gated cation channel. *Neuron* **17**, 681–693

22. Belluscio, L., Gold, G.H., Nemes, A. & Axel, R. (1998) Mice deficient in G(olf) are anosmic. *Neuron* **20**, 69–81

23. Kinnamon, S.C. & Margolskee, R.F. (1996) Mechanisms of taste transduction. *Curr. Opin. Neurobiol.* **6**, 506–513

24. Boekhoff, I., Tareilus, E., Strotmann, J. & Breer, H. (1990) Rapid activation of alternative second messenger pathways in olfactory cilia from rats by different odorants. *EMBO J.* **9**, 2453–2458

25. Raming, K., Krieger, J., Strotmann, J., Boekhoff, I., Kubick, S., Baumstark, C. & Breer, H. (1993) Cloning and expression of odorant receptors. *Nature (London)* **361**, 353–356

26. Sengupta, P., Chou, J.H. & Bargmann, C.I. (1996) odr-10 encodes a seven transmembrane domain olfactory receptor required for responses to the odorant diacetyl. *Cell* **84**, 899–909

27. Zhang, Y., Chou, J.H., Bradley, J., Bargmann, C.I. & Zinn, K. (1997) The *Caenorhabditis elegans* seven-transmembrane protein ODR-10 functions as an odorant receptor in mammalian cells. *Proc. Natl. Acad. Sci. U.S.A.* **94**, 12162–12167

28. Gat, U., Nekrasova, E., Lancet, D. & Natochin, M. (1994) Olfactory receptor proteins. Expression, characterization and partial purification. *Eur. J. Biochem.* **225**, 1157–1168

29. Kiefer, H., Krieger, J., Olszewski, J.D., Von Heijne, G., Prestwich, G.D. & Breer, H. (1996) Expression of an olfactory receptor in *Escherichia coli*: purification, reconstitution, and ligand binding. *Biochemistry* **35**, 16077–16084

30. Wellerdieck, C., Oles, M., Pott, L., Korsching, S., Gisselmann, G. & Hatt, H. (1997) Functional expression of odorant receptors of the zebrafish *Danio rerio* and of the nematode *C. elegans* in HEK293 cells. *Chem. Senses* **22**, 467–476

31. Zhao, H., Ivic, L., Otaki, J.M., Hashimoto, M., Mikoshiba, K. & Firestein, S. (1998) Functional expression of a mammalian odorant receptor. *Science* **279**, 237–242

32. Ressler, K.J., Sullivan, S.L. & Buck, L.B. (1993) A zonal organization of odorant receptor gene expression in the olfactory epithelium. *Cell* **73**, 597–609

33. Vassar, R., Ngai, J. & Axel, R. (1993) Spatial segregation of odorant receptor expression in the mammalian olfactory epithelium. *Cell* **74**, 309–318

34. Lancet, D. (1991) The strong scent of success. *Nature (London)* **351**, 275–276

35. Chess, A., Simon, I., Cedar, H. & Axel, R. (1994) Allelic inactivation regulates olfactory receptor gene expression. *Cell* **78**, 823–834

36. Troemel, E.R., Kimmel, B.E. & Bargmann, C.I. (1997) Reprogramming chemotaxis responses: sensory neurons define olfactory preferences in *C. elegans*. *Cell* **91**, 161–169

37. Qasba, P. & Reed, R.R. (1998) Tissue and zonal-specific expression of an olfactory receptor
     transgene. *J. Neurosci.* **18**, 227–236
38. Lancet, D., Greer, C.A., Kauer, J.S. & Shepherd, G.M. (1982) Mapping of odor-related neuronal
     activity in the olfactory bulb by high-resolution 2-deoxyglucose autoradiography. *Proc. Natl. Acad.
     Sci. U.S.A.* **79**, 670–674
39. Mombaerts, P., Wang, F., Dulac, C., Chao, S.K., Nemes, A., Mendelsohn, M., Edmondson, J. &
     Axel, R. (1996) Visualizing an olfactory sensory map. *Cell* **87**, 675–686
40. Singer, M.S., Shepherd, G.M. & Greer, C.A. (1995) Olfactory receptors guide axons. *Nature
     (London)* **377**, 19–20
41. Mori, K., Mataga, N. & Imamura, K. (1992) Differential specificities of single mitral cells in rabbit
     olfactory bulb for a homologous series of fatty acid odor molecules. *J. Neurophysiol.* **67**, 786–789

9

Pathology and drug action in schizophrenia: insights from molecular biology

Philip G. Strange

School of Animal and Microbial Sciences, University of Reading,
Whiteknights, Reading RG6 6AJ, U.K.

Introduction

The application of molecular biology techniques to the study of the brain has provided major new insights into diseases of the brain and their treatment, as well as into normal brain function. This essay illustrates this impact of molecular biology by considering one brain disorder, schizophrenia, and its treatment.

Schizophrenia: the clinical picture

Schizophrenia is a severe disorder of personality that afflicts about 1% of the population [1]. Patients experience a variety of symptoms that have been divided into subgroups of positive symptoms (e.g. thought disorder, abnormal beliefs and experiences) and negative symptoms (e.g. poverty of speech, loss of emotional response, reduced motor function). This division may be helpful in relation to the way patients present, but more recently the syndrome of schizophrenia has been described as having three dimensions [2,3]. In this categorization there are the negative symptoms as before, but the positive symptoms are divided into further subgroups of psychotic and disorganized. It is thought that these three dimensions of symptoms may reflect different disease processes, emphasizing that schizophrenia may not be a homogeneous disorder.

Schizophrenia has a genetic basis, and evidence for this comes from the incidence of schizophrenia in families and in twin pairs. There is debate as to how great the role of genetic factors is; the current consensus is that it is quite high, but is probably mediated by several genes, with a minor but nevertheless important role for environmental factors [4].

There have been many theories regarding the occurrence of schizophrenia. Here I consider changes in neurotransmitters and in the structure of the brain, and how these changes might relate to the causes of schizophrenia.

Changes in the brain in schizophrenia

Changes in neurotransmitter systems in the brain

Until relatively recently it had been difficult to provide convincing evidence for structural changes in the brains of those suffering from schizophrenia. Theories of schizophrenia were proposed, therefore, based on changes in neurotransmitter systems in the brain. In order to investigate these ideas, many neurotransmitters and their metabolites were measured in *post mortem* brains from patients who had suffered from schizophrenia in life, but no consistent alterations have been described.

One neurotransmitter that has been a particular focus of attention in this context is dopamine. Receptors for this neurotransmitter (Figure 1) have been actively examined owing to the prominent actions at dopamine receptors of the drugs used to treat the disorder (anti-psychotic drugs). Ligand-binding studies measuring D_1-like and D_2-like dopamine receptors in *post mortem* brains from schizophrenics showed no changes in the D_1-like subgroup, but in several laboratories an increased number of D_2-like receptors was reported [5]. Although some studies reported an increased receptor number in drug-naive patients, in many cases the patients used for the studies had received anti-psychotic drugs in life, and it is known that such treatment raises the numbers of D_2-like receptors in the brain. Much of the reported elevation in D_2-like receptor number in schizophrenia may therefore be attributed to the drug treatments, but there are reports that there may be an elevation in receptor number that is independent of drug treatment in a subgroup of patients [5]. More recently, following the cloning of five dopamine receptor subtypes (see Figure 1), it has been reported that there is an increase in the numbers of D_4-dopamine receptors in the brains of schizophrenics, but these findings have not been replicated [6].

These studies on the numbers of D_2-like receptors in *post mortem* brains measured using ligand binding have been complemented by studies using *in vivo* imaging techniques [positron emission tomography (PET) with ^{11}C-labelled ligands], but unfortunately these studies have also been contradictory. Some studies report an elevation in receptor numbers in drug-naive patients, but in other studies these findings have not been replicated [7]. It may be that a problem with the studies is the heterogeneous nature of the patient population.

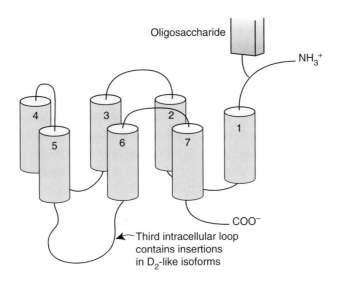

Figure 1. Multiple dopamine receptor subtypes

Work on receptors for the neurotransmitter dopamine using physiological, pharmacological and biochemical techniques suggested that there were two dopamine receptors (D_1 and D_2) with different biochemical and pharmacological properties and different functions. Molecular biological techniques, however, have defined five dopamine receptor sequences (D_1, D_2, D_3, D_4 and D_5). Each of these is predicted to be a G-protein-coupled receptor with seven membrane-spanning α-helical regions linked by intracellular and extracellular loops and folded together as shown in the diagram. The amino acid sequences of these receptors are related, but comparisons allow the receptors to be divided into two subgroups (D_1, D_5 and D_2, D_3, D_4). The pharmacological properties of the two subgroups resembled respectively the D_1 and D_2 receptors defined using biochemical and pharmacological techniques, and so the two subgroups are often referred to as D_1-like (D_1, D_5) and D_2-like (D_2, D_3, D_4). The complexity of the D_2-like receptors is increased by isoforms differing in the size (short/long) of the third intracellular loop for the D_2 and D_3 receptors and by polymorphisms of the D_4 receptor in the human population [37]. In this article dopamine receptor subtypes defined using molecular biological criteria are designated D_1–D_5, but are described as D_1-like or D_2-like when the subtype has been defined only by pharmacological properties.

A major goal of this work has been to verify 'dopamine hypotheses' of schizophrenia, which assert that the disorder is due to an elevation in dopamine function in the brain. Elevated dopamine function could occur at the level of dopamine receptors or at the level of dopamine itself. As outlined above, there is little consistent evidence that dopamine receptors are increased in the brain in schizophrenia. Studies on dopamine and its metabolites have been no more illuminating; changes have been described in some reports, but these have not been replicated [8]. More recently, however, it has been reported that the dopamine biosynthetic enzyme, dopa decarboxylase, is increased in schizophrenia [9]. Using *in vivo* imaging (PET) it has been shown that, in schizophrenia, there is an increased ability of the drug amphetamine to cause dopamine release [10]. Also, it has been shown that the impaired cortical activation seen in patients with schizophrenia when performing a verbal fluency task can be reversed by administration of the dopamine agonist apomorphine [11]. There is therefore continuing interest in 'dopamine hypotheses' of schizophrenia.

Many other neurotransmitter systems have been examined in respect of schizophrenia, but no consistent pattern emerges. Glutamate systems have been studied and theories of schizophrenia have been proposed based on reduced glutamate activity [12]. Another neurotransmitter that has attracted interest is 5-hydroxytryptamine (5-HT; serotonin) [13], and there are reports showing that the numbers of 5-HT receptors of the 5-HT$_{2A}$ class are lower in brains from schizophrenics. There have also been reports that a common allelic polymorphism (T102C) of this receptor is associated with an increased risk of schizophrenia, but these findings have not been replicated in all studies [14]. Interactions of both the glutamate and 5-HT systems with dopamine systems have been proposed in relation to the symptoms and therapies of schizophrenia [13,14].

Changes in brain structure

The use of modern imaging techniques for the brain in life [computerized axial tomography (CAT) and magnetic resonance imaging (MRI)] and the application of careful morphological studies to *post mortem* brains have shown that there may be small but consistent changes in the structure of the brain in schizophrenia [15]. The most robust observation is an increased lateral ventricular size and a reduced temporal lobe size. These changes do not seem to be progressive in nature and seem to be independent of the symptoms (positive, negative) exhibited by the patient, and so they may reflect a neurodevelopmental alteration [3,4,16]. There may also be changes in the functions of the frontal lobes, including reduced activation during the performance of certain tasks [15]. These findings have been supported by demonstrations of histological differences between the brains of schizophrenics and normal individuals, e.g. hippocampal pyramidal cell disarray, which would also imply changes in the development of the brain.

It is thought, therefore, that schizophrenia results from a disturbance in the normal development of the brain and that the symptoms arise from the disturbed brain function that results. Brain development may be altered by a variety of influences, including minor brain damage at birth, or by genetically programmed changes in development. Changes in the functions of dopamine, 5-HT or glutamate systems (see above) would then be secondary to these changes in development.

Genetic linkage analysis of schizophrenia

There have been extensive attempts to demonstrate a role for particular genes in the incidence of schizophrenia. With the discovery of the different dopamine receptor subtypes using molecular biological techniques (see Figure 1), and because of the interest in the role of dopamine receptors in schizophrenia, attempts have been made to demonstrate linkage between the occurrence of schizophrenia and the receptor genes; however, despite some reports claiming such a linkage, these have not been confirmed subsequently [17].

Some studies have been performed attempting to link the disorder to markers on different chromosomes (e.g. chromosomes 5 and 22). Although some of these reports were greeted with enthusiasm, again most have not been replicated. The most convincing linkage report so far is between a marker on chromosome 6 and the disorder, which has been observed in several independent studies [18]. Further extensive work will be required in order to understand the significance of this observation, but it provides the first clear breakthrough in understanding the genetics and hence the biology of schizophrenia.

Another area where genetic linkage analysis may be important is in susceptibility to drug treatment. For example, it could be that a mutation in a receptor gene might alter the susceptibility of an individual to the effects a drug. Indeed, it has been claimed that there is an association between the T102C allele of the 5-HT$_{2A}$ receptor and the lack of response to the drug clozapine. These findings have not, however, been replicated [13,14].

Drug action in schizophrenia

Typical and atypical anti-psychotics

A large number of drugs (anti-psychotic drugs) exist that have been or are used to treat schizophrenia (e.g. chlorpromazine, haloperidol), and the most prominent therapeutic effect of these drugs is to reduce the incidence of the positive symptoms of the disorder [1]. Some patients are, however, resistant to the therapeutic effects of the drugs. Some drugs (e.g. clozapine) have been reported to have effects on negative symptoms. In addition, many of the drugs induce motor side-effects (extrapyramidal side-effects), most notably a Parkinsonian-like syndrome in some patients after a short period of treatment and, after longer-term treatment, a hyperkinetic disorder, tardive dyskinesia.

Anti-psychotic drugs have been divided into two groups, the so-called typical and atypical anti-psychotics. Typical anti-psychotics (e.g. haloperidol) are defined as drugs that are therapeutically effective against the positive symptoms of schizophrenia and which also induce some of the side-effects outlined above. The atypical drugs (e.g. clozapine and risperidone) are defined as those drugs with therapeutic effects against the positive symptoms of schizophrenia but with a reduced tendency to induce motor side-effects. There has been some debate about this definition [19,20] as to whether the definition of atypical should include either efficacy against negative symptoms or efficacy in treatment-resistant patients. The simple definition is still widely used, and some feel that the differences between typical and atypical drugs are quantitative rather than qualitative.

It is of some interest to try to understand the basis of the actions of these drugs and the biochemical basis of the distinction between the typical drugs and the atypical drugs. This is an area to which molecular biological studies have contributed significantly.

Dopamine and 5-HT receptors as drug targets

Following the discovery of dopamine as a neurotransmitter in the brain, it was shown that the anti-psychotic drugs were somehow interfering with the actions of dopamine in the brain and, based on behavioural studies, this appeared to be via blockade of dopamine receptors [21]. With the advent of assays for the different dopamine receptor subtypes (stimulation of adenylate cyclase for D_1-like receptors and ligand-binding assays for D_2-like receptors), it became clear that there was an excellent correlation between the daily dose of a range of drugs used to treat schizophrenia and the affinities of these drugs for D_2-like dopamine receptors [1]. The correlations reported are remarkable given the crudity of the measure of the activity of the drugs *in vivo*, but no similar correlation was found for any other measure over such a wide range of drugs. Thus it appears that an ability to bind to D_2-like dopamine receptors is an important determinant of a potential anti-psychotic drug. The anti-psychotic effects of these drugs are thought to be mediated via D_2-like receptors in limbic and cortical brain regions, as these brain regions are considered to be important in mediating the kinds of functions that are disturbed in schizophrenia. The motor (extrapyramidal) side-effects are thought to be mediated via D_2-like receptors in striatal regions. With the discovery of the different dopamine receptor subtypes using molecular biological studies (see Figure 1), it became clear that these D_2-like effects could be at D_2, D_3 or D_4 receptors, and the preferential localization of the D_3 and D_4 receptors in limbic and cortical brain regions made these attractive potential targets for anti-psychotic action. A drug that acted selectively on D_3 or D_4 receptors might be an anti-psychotic but have reduced extrapyramidal side-effects.

One way to determine which of the new dopamine receptor subtypes is responsible for anti-psychotic action would be to demonstrate preferential action of anti-psychotic drugs on a particular subtype. This can be done by comparing the dissociation constants of different drugs for the receptor subtypes using ligand-binding assays under carefully controlled conditions [22]. For some of the clinically useful drugs (e.g. haloperidol) there is little difference in their affinities for the three subtypes (Table 1). The D_4 receptor shows a markedly lower affinity for certain of the substituted benzamide anti-psychotics (e.g. raclopride, sulpiride), so that it cannot be the common target for anti-psychotic action. Based on these observations it seems that anti-psychotic action may include occupancy of D_2 and D_3 receptors, but D_4 receptor occupancy is not mandatory. There may, however, be significant mechanistic differences between drugs, for example actions at other receptors, but these analyses of actions at dopamine receptor subtypes do not account for the typical/atypical distinction. The typical/atypical definition is, however, a rather loose definition based on overall clinical efficacy, so different drugs may be atypical for different mechanistic reasons.

Clozapine is an atypical anti-psychotic drug that has generated much interest owing to its therapeutic effects in patients resistant to other anti-psychotics and its effects on negative symptoms, and it is important to try to understand the mechanism of these effects. Much has been made of the higher affinity of this drug for the D_4 dopamine receptor compared with the D_2 receptor [23], and it has been argued that actions at D_4 receptors may contribute to the atypical actions of this drug. There has been some discussion about the exact affinity of clozapine for the D_4 dopamine receptor [22–24], but, based on carefully controlled experiments, an affinity of approx. 7 nM can be determined (Table 1) and the D_2/D_4 preference is about 5-fold [22]. The higher affinity of the D_4 receptor for clozapine could mean that it is preferentially occupied when clozapine is used therapeutically, which would render clozapine different mechanistically from many other anti-psychotics. It is, however, difficult to define the actual concentrations of drugs at receptor sites in the brain, so it is difficult to know if this D_2/D_4 preference is actually reflected in differential occupancies. It is more likely that actions of clozapine at other receptors (e.g. 5-HT receptors; see below) are responsible for the unusual actions of this drug, but there is considerable debate about this.

The recognition of the potential role of the D_3 and D_4 receptors as novel sites of anti-psychotic action has led to intensive efforts to synthesize selective

Table 1. Dissociation constants of anti-psychotic drugs at dopamine and 5-HT receptor subtypes

Values for the D_2(short) and D_3 receptors were taken from [38], and were obtained in competition experiments versus [^{3}H]raclopride binding to the receptors expressed in CHO cells. [^{3}H]Raclopride is a radioligand that can be used under conditions that avoid experimental artefacts [22]. For the D_4 receptor the data were taken from [39], and were obtained using competition versus [^{3}H]spiperone binding corrected for potential artefacts [22]. The data for the D_4 receptor must, therefore, be taken as estimates of affinity. Data for the $5HT_{2A}$ receptor are from [40].

	Dissociation constant (nM)			
	D_2	D_3	D_4	$5HT_{2A}$
Chlorpromazine	0.55	1.2	11.6	2
Clozapine	35	83	7	3.8
Haloperidol	0.53	2.7	0.78	58
Raclopride	1	1.8	800	4400
Remoxipride	54	969	930	6400
Risperidone	1.3	6.7	2.3	0.20
Sertindole	0.38	1.63	–	0.29
(−)-Sulpiride	2.5	8	333	–
Thioridazine	1.2	2.3	–	1.3

compounds for these receptors, the first of which are being tested in animal models [25,26]. L745870, a D_4-selective compound, was tested as an anti-psychotic in humans, but was ineffective [26], supporting the view that activity at the D_4 receptor is not a prerequisite for anti-psychotic activity.

Many anti-psychotics have also been shown to have significant affinities for 5-HT receptors, e.g. the 5-HT$_2$ receptor. It is now clear, based on molecular biological studies, that there are a large number of 5-HT receptor subtypes [13]. These have been the subject of genetic studies similar to those outlined above. In addition, however, it has been shown that some of the atypical drugs have a high affinity for some of these 5-HT receptor subtypes. In some cases the affinity is higher than for the D_2 receptor, and this could contribute to their unusual therapeutic profile.

The 5-HT$_{2A}$ receptor is of some interest here. The drug ritanserin has a high affinity at this receptor and a rather low affinity at dopamine receptors, and when administered together with typical anti-psychotic drugs it has some efficacy against negative symptoms [27]. It has also been suggested that atypical anti-psychotics can be distinguished more generally by their higher affinity for 5-HT$_2$ receptors relative to D_2 receptors [28]. Actions of drugs at 5-HT$_{2A}$ receptors could suppress both negative symptoms and extrapyramidal side-effects via indirect effects on dopamine activity in cortical/limbic and striatal brain regions respectively [13]. Table 1 gives some values for the dissociation constants of typical (e.g. haloperidol, chlorpromazine) and atypical (clozapine, remoxipride, sertindole) anti-psychotics at 5-HT$_{2A}$ receptors. The recently described 5-HT$_6$ and 5-HT$_7$ receptors also have high affinities for several atypical drugs, including clozapine, and so may contribute to their actions [13]. As indicated above, however, different drugs may be atypical through different mechanisms, and for clozapine this is very difficult to define owing to its actions at a wide variety of sites in addition to dopamine and 5-HT receptors.

In vivo occupancy of dopamine and 5-HT receptors

Determination of the dissociation constants of drugs for binding to different receptors can give indications of *in vivo* drug occupancies. However, owing to the problems in defining the concentrations of drugs at receptor sites in the brain, these analyses can be misleading, and it is desirable to determine drug occupancies more directly. The development of receptor-selective labelled ligands has enabled drug occupancy of receptors in the brains of living individuals to be examined using PET and single-photon emission tomography scanning. For the dopamine receptors, probes that are currently available are only able to distinguish D_1-like and D_2-like receptors. More selective ligands will be required in order to analyse the individual receptor subtypes.

PET scans with the probe [^{11}C]raclopride have been used to determine the occupancies of D_2-like receptors during anti-psychotic therapy. These studies have shown that, at typical doses, anti-psychotic drugs such as haloperidol occupy 70–90% of the D_2-like receptors in human brain and that this occupancy occurs quickly [29]. Occupancy of D_1-like receptors by many of the drugs was much lower (0–44%). Motor side-effects were seen in patients when D_2-like receptor occupancy exceeded ~80%. It was also found that haloperidol could be used successfully at lower doses to treat patients, and under these conditions occupancy of D_2-like receptors was only about 50%; motor side-effects were absent, which fits well with the threshold for occurrence of these side-effects suggested above [30].

The atypical drug clozapine has been studied using these techniques [7], and it has been found that at normal doses it occupies 38–63% of the D_2-like receptors in the striatum and 38–52% of the D_1-like receptors. These occupancies are consistent with the affinities of clozapine for the two subclasses of receptors when compared with the affinities of the typical drugs, although it is in fact very difficult to know the true concentrations of drugs at receptor sites in the brain. Clozapine, for example, is concentrated in the brain more than 20-fold over the serum, and other drugs are concentrated to different extents [31]. In one study, however, *in vivo* receptor occupancies in the striatum with increasing doses of clozapine were examined [29]; whereas 5-HT$_2$ receptors were occupied to a level of 96%, the maximum occupancy of D_2-like receptors was 61%, and for D_1-like receptors this figure was 48%. In complementary experiments raclopride was shown to occupy 97% of the D_2-like receptors. Therefore, even with apparently saturating concentrations of drug, clozapine is unable to occupy all the D_2-like receptors. These observations are difficult to explain, given that in *in vitro* assays clozapine and raclopride are fully competitive and seem to label the same set of D_2-like receptor sites.

It is beginning to be possible to examine extra-striatal D_2-like receptors using high-affinity probes such as [123I]epidepride, and this has been used to examine D_2-like receptors in the temporal cortex of schizophrenics [32]. Surprisingly, clozapine was able to occupy ~90% of the D_2-like receptors in this brain region. These results were interpreted in terms of a limbic selectivity for the drug, but further controls are required to confirm these results. In particular, it will be important to define whether in cortical regions [123I]epidepride is labelling other sites in addition to the D_2-like receptors (see [33]).

Inverse agonism of anti-psychotic drugs

It has been widely assumed that the mechanism of action of the anti-psychotic drugs is simply to act as antagonists of dopamine at its receptors (D_2-like). This notion has recently been challenged through the demonstration that a wide range of anti-psychotic drugs (typical and atypical) act as inverse agonists at D_2 receptors expressed in CHO cells [34]. Inverse agonism refers to the ability of these drugs to act in an opposite manner to agonists; whereas agonists increase the activity of a receptor system, inverse agonists suppress the activity of a receptor system. Similar, although less extensive, data describing inverse agonism were reported for D_2 receptors in pituitary cells [35]. These findings have significant implications for understanding anti-psychotic drug action. Firstly, in the studies on D_2 receptors in CHO cells all of the anti-psychotic drugs exhibited the same degree of inverse agonist efficacy. An aminotetralin UH-232 behaved as a neutral antagonist, but this drug has not been tested as an anti-psychotic. It is therefore unclear whether inverse agonist efficacy is essential for anti-psychotic action. If it is important, then the possibility arises that different degrees of inverse agonist efficacy may be built into drugs to provide optimum activity. Secondly, the D_2 receptors in the CHO cells exhibited constitutive (agonist-independent) activity and this was reversed by the anti-psychotics. This raises the possibility that there is constitutive activation of D_2 receptors in normal brain, although this has yet to be demonstrated. Constitutive activation could vary in different brain regions and this could provide for a regional selectivity in drug action. Constitutive activation could also underlie certain disease states. Thirdly, even if there were no constitutive activation in normal brain, the demonstration that these drugs are inverse agonists rather than antagonists may be important. It has been shown that treatment of humans or experimental animals with anti-psychotic drugs increases the number of D_2-like receptors in the brain [5]. This had been assumed to be due to the blockade of access of dopamine to the receptors, but it may be a reflection of the inverse agonism of the drugs, agonists having been generally found to reduce receptor number. The effects of anti-psychotic drugs in increasing receptor numbers in cell lines expressing D_2 receptors are in agreement with these observations [36].

Conclusions

Molecular biological studies have contributed significantly to the understanding of both the causes and the treatments of schizophrenia. With respect to the causes of schizophrenia, the main contribution is towards defining genetic factors that may influence the occurrence of the disorder. There is much work to be done here, but this should lead ultimately to an understanding of the causes of the disorder. This may also result in genetic tests for susceptibility to schizophrenia, and the ethical and moral implications of these will be great. With respect to treatments of schizophrenia, the main contribution is in understanding the actions of drugs at different receptor subtypes which themselves have been defined through gene cloning. Also, although this chapter has concentrated on one disorder, schizophrenia, the methods used for understanding pathology and treatments are applicable to other brain disorders.

Summary

- *Schizophrenia is a severe disorder of personality which has a genetic basis.*
- *Schizophrenia arises from a change in brain development.*
- *There is no strong evidence that disturbances in neurotransmitter systems are a primary cause.*
- *Anti-psychotic drugs act primarily through D_2 and D_3 dopamine receptors*
- *The atypical drug clozapine may act through a number of different receptors, including D_2, D_3 and D_4 dopamine receptors.*
- *Anti-psychotic drugs are inverse agonists at D_2 dopamine receptors.*

References

1. Strange, P.G. (1992) *Brain Biochemistry and Brain Disorders*, Oxford University Press, Oxford
2. Liddle, P.F. (1987) The symptoms of chronic schizophrenia: a re-examination of the positive–negative dichotomy. *Br. J. Psychiatry* **151**, 145–151
3. Andreasen, N.C. (1994) The mechanism of schizophrenia. *Curr. Opin. Neurobiol.* **4**, 245–251
4. Cannon, M. & Jones, P. (1996) Schizophrenia. *J. Neurol. Neurosurg. Psychiatry* **61**, 604–613
5. Seeman, P. & Niznik, H.B. (1990) Dopamine receptors and transporters in Parkinson's disease and schizophrenia. *FASEB J.* **4**, 2737–2744
6. Reynolds, G.P. (1996) The importance of dopamine D_4 receptors in the action and development of antipsychotic drugs. *Drugs* **51**, 7–11
7. Sedvall, G. & Farde, L. (1995) Chemical brain anatomy in schizophrenia. *Lancet* **346**, 743–749
8. Davis, K.L., Kahn, R.S., Ko, G. & Davidson, M. (1991) Dopamine in schizophrenia: a review and reconceptualization. *Am. J. Psychiatry* **148**, 1474–1486
9. Reith, J., Benkelfat, C., Sherwin, A., Yasuhan, Y., Kuwabara, H., Andermann, F., Bachneff, S., Cumming, P., Diksic, M., Dyre, S.E., et al. (1994) Elevated dopa decarboxylase activity in living brains of patients with psychosis. *Proc. Natl. Acad. Sci. U.S.A.* **91**, 11651–11654
10. Laruelle, M., Abi-Darghan, A., Van Dyck, C.H., Gil, R., D'Souza, C.D., Erdos, J., McCance, E., Rosenblatt, W., Fingado, C., Zoghi, S.S., et al. (1996) Single photon emission computerized tomography imaging of amphetamine induced dopamine release in drug free schizophrenic subjects. *Proc. Natl. Acad. Sci. U.S.A.* **93**, 9235–9240
11. Fletcher, P.C., Frith, C.D., Grasby, P.M., Friston, K.J. & Dolan, R.J. (1996) Local and distributed effects of apomorphine on fronto-temporal function in acute unmedicated schizophrenia. *J. Neurosci.* **16**, 7055–7062
12. Olney, J.W. & Farber, N.B. (1995) Glutamate receptor dysfunction in schizophrenia. *Arch. Gen. Psychiatry* **52**, 998–1007
13. Busatto, G.F. & Kerwin, R.W. (1997) Perspectives on the role of serotonergic mechanisms in the pharmacology of schizophrenia. *J. Psychopharmacol.* **11**, 3–12
14. Harrison, P.J. & Burnet, P.W.J. (1997) The 5-HT2A receptor gene in the aetiology, pathophysiology and therapy of schizophrenia. *J. Psychopharmacol.* **11**, 18–20
15. Chua, S.E. & McKenna, P.J. (1995) Schizophrenia – a brain disease. *Br. J. Psychiatry* **166**, 563–582
16. Ross, C.A. & Pearlson, G.D. (1996) Schizophrenia, the heteromodal association cortex and development: potential for a neurogenetic approach. *Trends Neurosci.* **19** 171–176
17. Coon, H., Byerley, W., Holik, J., Hoff, M., Myels-Worsley, M., Lannfelt, L., Sokoloff, P., Schwartz, J.C., Waldo, M., Freedman, R. & Plaetke, R. (1993) Linkage analysis of schizophrenia with five dopamine receptor genes in nine pedigrees. *Am. J. Hum. Genet.* **52**, 327–334

18. Peltonen, L. (1995) All out for chromosome six. *Nature (London)* **378**, 665–666
19. Waddington, J.L. & O'Callaghan, E. (1997) What makes an antipsychotic atypical? Conserving the definition. *CNS Drugs* **7**, 341–346
20. Kane, J.M. (1997) What makes an antipsychotic atypical? Should the definition be preserved? *CNS Drugs* **7**, 347–348
21. Leysen, J.E. & Niemegeers, C.J.E. (1985) Neuroleptics. *Handb. Neurochem.* **9**, 331–361
22. Strange, P.G. (1997) Commentary on atypical neuroleptics. *Neurospychopharmacology* **16**, 116–122
23. Seeman, P. (1992) Dopamine receptor sequences. Therapeutic levels of neuroleptics occupy D_2 receptors, clozapine occupies D_4 receptors. *Neuropsychopharmacology* **7**, 261–284
24. Van Tol, H.H.M., Wu, C.M., Guan, H.C., Ohara, K., Bunzow, J.R., Civelli, O., Kennedy, J., Seeman, P., Niznik, H.B. & Jovanovic, V. (1992) Multiple dopamine D_4 receptor variants in the human population. *Nature (London)* **358**, 149–152
25. Bolton, D., Boyfield, I., Coldwell, M.C., Hadley, M.S., Healy, M.A.M., Johnson, C.N., Markwell, R.E., Nash, D.J., Riley, G.J., Stemp, G. & Wadsworth, H.J. (1996) Novel 2,5-disubstituted-1H pyrroles with high affinity for the dopamine D_3 receptor. *Bioorg. Med. Chem. Lett.* **11**, 1233–1236
26. Bristow, L.J., Kramer, M.S., Kulagowski, J., Patel, S., Regan, C.I. & Seabrook, G.R. (1997) Schizophrenia and L745870, a novel dopamine D_4 receptor antagonist. *Trends Pharmacol. Sci.* **18**, 186–188
27. Duinkerke, S.J., Botter, P.A., Jansen, A.A.I., Van Dongen, P.A.M., Van Haafter, A.J., Boom, A.J., Van Laarhoven, J.H.M. & Busard, H.L.S.M. (1993) Ritanserin, a selective 5-HT2/1C antagonist and negative symptoms in schizophrenia. *Br. J. Psychiatry* **163**, 451–455
28. Meltzer, H.B. (1990) The importance of serotonin–dopamine interactions in the actions of clozapine. *Br. J. Psychiatry* **160** (Suppl. 17), 22–29
29. Nordstrom, A., Farde, L., Nyberg, S., Karlsson, P., Halldin, C. & Sedvall, G. (1995) D1, D2 and 5HT2 receptor occupancy in relation to clozapine serum concentration. A PET study of schizophrenic patients. *Am. J. Psychiatry* **152**, 1444–1449
30. Nyberg, S., Farde, L., Halldin, C., Dahl, M.L. & Bertilson, L. (1995) D_2 receptor occupancy during low dose treatment with haloperidol decanoate. *Am. J. Psychiatry* **152**, 173–178
31. Brunello, N., Masotto, C., Steardo, L., Markstein, R. & Racagni, G. (1995) New insights into the biology of schizophrenia through the mechanism of action of clozapine. *Neuropsychopharmacology* **13**, 177–213
32. Pilowsky, L., Mulligan, R.S., Acton, P.D., Eil, P.J., Costa, D.C. & Kerwin, R.W. (1997) Limbic selectivity of clozapine. *Lancet* **350**, 490–491
33. Joyce, J.N., Janowsky, A. & Neve, K.A. (1991) Characterization and distribution of [^{125}I]epide-pride binding to D_2 receptors in basal ganglia and cortex of human brain. *J. Pharmacol. Exp. Ther.* **257**, 1253–1263
34. Hall, D.A. & Strange, P.G. (1997) Evidence that antipsychotic drugs are inverse agonists at D_2 dopamine receptors. *Br. J. Pharmacol.* **121**, 731–736
35. Nilsson, C.L. & Eriksson, F. (1993) Haloperidol increases prolactin release and cAMP formation *in vitro* – inverse agonism at dopamine D_2 receptors. *J. Neural Transm.* **92**, 213–220
36. Boundy, V.A., Pacheco, M.A., Guan, W. & Molinoff, P.B. (1995) Agonists and antagonists differentially regulate the high affinity state of the D_{2L} receptor in human embryonic kidney cells. *Mol. Pharmacol.* **48**, 956–964
37. Neve, K.A. & Neve, R.L. (1997) *The Dopamine Receptors*, Humana Press, Totowa, NJ
38. Malmberg, A., Jackson, D.M., Eriksson, A. & Mohell, N. (1993) Unique binding characteristics of antipsychotic agents interacting with human dopamine D_{2A}, D_{2B} and D_3 receptors. *Mol. Pharmacol.* **43**, 749–754
39. Seeman, P. & Van Tol, H.H.M. (1994) Dopamine receptor pharmacology. *Trends Pharmacol. Sci.* **15**, 264–269
40. Seeman, P., Corbett, R. & Van Tol, H.H.M. (1997) Reply to commentaries. *Neuropsychopharmacology* **16**, 127–135

10

Genetics of Alzheimer's disease

Michael Hutton, Jordi Pérez-Tur and John Hardy[1]

Neurogenetics Laboratory, The Mayo Clinic Jacksonville, 4500 San Pablo Rd., Jacksonville, FL 32224, U.S.A.

Introduction

Before the application of molecular genetics to the study of Alzheimer's disease (AD) in the late 1980s, there were very many theories about the disease. These hypotheses were generally ill-formed, and often did not discriminate between theories concerning the aetiology of the disease, theories that addressed pathogenic mechanisms and theories that dealt with neuronal loss and treatment strategies. Indeed, these areas were dealt with almost in isolation and without reference to each other.

It was very difficult to break through the intellectual fog that surrounded AD, because there were very limited ways in which the disease could be studied. The main approach was to study the pathology of the disease; through this approach, a huge number of neurochemical and molecular changes in AD brains were documented. Unfortunately, of its nature, this approach is very poor at determining the order of changes in the brain: which changes came first or early in the disease process, and which followed as a result of the damage caused by the disease. This is extremely important, because it is our fundamental wish, once we have observed the mass of AD pathology, to put it in a rational order: to reduce, therefore, the disease pathology to a pathway akin to the glycolytic pathway, i.e. one of cause and effect. This is our aim, because when this is achieved and the pathogenic pathway is understood, treatment strategies can be designed to interfere in this process. Of course, it is likely that

[1]*To whom correspondence should be addressed.*

the pathway of the AD process will not be a simple linear process like the gly-colytic pathway, but rather it is more likely to be an inverted pyramid shape, with an increasing degree of complexity as more damage accrues from the initial insult. Thus it is likely that the initial events in the disease process will be relatively simple, but that, as one moves further from this initial insult, the complexity of the problem will become greater as the damage affects more molecules in more cellular compartments. This idea of increasing complexity is important, because it implies that treatment strategies are more likely to be completely successful if they are aimed close to the initiating insult.

The power and beauty of molecular genetics is its simplicity and the fact that no assumptions are made about disease aetiology except for the mode of inheritance. The experimental paradigm is very simple: in families with auto-somal dominant disease, one 'simply' tries to find the mutant gene that under-lies the disease in that family, since it will be shared by all affected family members but not by unaffected family members. In the study of autosomal dominant AD, a false start occurred when the assumption was made that all cases of this form of the disease had mutations in the same gene [1]. However, since it was realized that more than one gene is involved in early-onset AD, progress has been fairly swift, and three genes have been implicated in early-onset AD and one locus as a risk factor for the disease.

In this essay, we shall show how an understanding of the genetics of AD has allowed the dissection of the early stages of disease pathogenesis, and we shall describe the likely nature of this process. We shall also briefly discuss whether there are likely to be other genes involved in this form of the disease.

Early-onset, autosomal dominant disease: the amyloid precursor protein (APP) and the presenilins

To date, three loci have been associated with early-onset autosomal AD; the first to be identified, on chromosome 21, was APP [2]. Mutations in APP were identified in a series of large families with autosomal dominant early-onset AD, with an age of onset of 45–65 years [3]. APP is processed by cellular proteases known as α-, β- and γ-secretases. The α-secretase cleaves within the Aβ peptide sequence in APP, preventing the generation of Aβ. In contrast, the β-secretase cleaves at the N-terminus and leads to the generation of the Aβ peptide, after cleavage by a γ-secretase. The Aβ peptide is the primary component of amyloid deposited in the brains of AD patients. All of the mutations found in APP occur either within or flanking the Aβ peptide sequence, and all affect the processing of APP to Aβ, such that either more total Aβ or more of the amyloidogenic Aβ42(43) species [Aβ variant containing 42 or 43 residues, compared with the normal variant containing 40 residues (Aβ40)] alone is generated. This longer form of Aβ is deposited selectively in the early stages of amyloid deposition.

These observations led to the hypothesis that the deposition of amyloid, made up principally of Aβ42(43), was the central event in the initial pathogenesis of AD: the 'amyloid cascade hypothesis'. This hypothesis was implicit in the work of those who first characterized the amyloid from the brains of AD patients [4,5]. Thus this genetic work supported the notion that the amyloid deposits in AD were not a consequence of the disease process, but rather were close to the cause of the disease.

However, mutations in APP are very rare, and to date have only been identified in about 20 families with autosomal dominant AD (for a review, see [3]). Most families do not have APP mutations, and it was clearly of importance not only to find the genes involved in other forms of autosomal dominant AD, but also to determine whether they act via an 'amyloidogenic' mechanism.

With the positional cloning of the presenilin-1 (PS-1) gene on chromosome 14, the major locus for autosomal dominant AD was discovered [6]. The age of onset in families with PS-1 mutations is the earliest observed (28–60 years) [7]. Database comparisons with the PS-1 sequence identified two expressed sequence tags with significant identity, suggesting the presence of a second related gene, presenilin-2 (PS-2) [8–10]. The subsequent identification of a mutation in PS-2 in a series of German families originally from the Volga valley in Russia confirmed the significance of this gene in AD pathogenesis [8–10]. Mutations identified in the PS-2 gene to date appear to result in AD with a later and more variable age of onset than mutations in PS-1 [8–10].

The structures of the PS-1 and PS-2 genes are remarkably similar. Each gene consists of a total of 13 exons, with 10 exons (exons 3–12) comprising the coding sequence; the 5' untranslated region is contained on four separate exons (1a, 1b, 2 and 3) [9,11–13]. In the PS-1 gene, exons 1a and 1b (in this article, for the sake of simplicity, we have kept to our original numbering system for the presenilin exons, which predated the discovery of the most 5' exon [9,11]) are mutually exclusive and represent alternative transcriptional start sites. The PS-1 promoter contains consensus TATA and CAAT box sequences; in addition, multiple STAT elements are observed that are involved in transcriptional activation in response to signal transduction [13]. The PS-2 promoter, in contrast, lacks consensus TATA and CAAT boxes, but contains regions of GC-rich sequence upstream of two transcriptional start sites. The equivalent positions of intron/exon boundaries in the coding sequences of the two genes are also virtually identical, consistent with the idea that they are derived from a common ancestral gene that underwent gene duplication (Figure 1).

Both presenilin genes undergo alternative splicing; however, the pattern is different in PS-1 and PS-2. In PS-1, alternative use of the splice donor site for exon 3 results in the inclusion/exclusion of codons 26–29 (Val-Arg-Ser-Gln; VRSQ) [9]. The inclusion of the VRSQ motif creates potential phosphorylation sites for protein kinase C and casein kinase II. However, the relative levels of the two transcripts do not appear to vary markedly in all tissues examined, suggesting that there is little tissue-specific regulation of this alternative splice

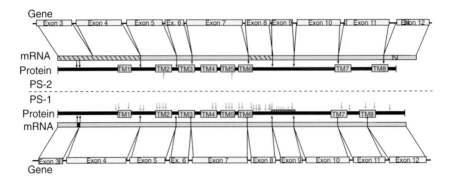

Figure I. Structures of the PS-I and PS-2 genes showing the similarity between them
Blue arrows indicate the positions of known mutations, including the Δ9 mutation. Cross-hatching indicates the alternative splicing that occurs in PS-2. TM1, TM2 etc. refer to transmembrane domains (see Figure2).

event [7,9]. In PS-2, the VRSQ motif is only partially conserved (WRSQ) and, as a result, is not alternatively spliced [11]. A second splice variant in PS-1 results in the presence or absence of exon 8, although this has only been observed in leucocytes [10,11]. The importance of this variant is unclear. However, it may be significant that exon 8 contains the highest density of AD mutations [14] and the equivalent exon is also alternatively spliced in PS-2 [10–12]. A further variant in PS-2 results in the alternative splicing of exons 4 and 5 [11]; the absence of these exons results in the loss of the normal translational start site. Transcripts lacking exons 4 and 5 make truncated proteins in transfected cells, presumably using an alternative downstream AUG codon (C. Haass, M. Hutton, J. Pérez-Tur and J. Hardy, unpublished work).

Two major transcripts are observed for both PS-1 and PS-2 on Northern blots. PS-1 mRNAs are approx. 2.7 and 7.5 kb in size, with the longer transcript thought to be generated by alternative polyadenylation site usage. PS-2 mRNAs are 2.3 and 2.6 kb in size. The PS-1 and PS-2 mRNAs encode proteins that display 67% identity to each other. The structure of the presenilin proteins (Figure 2) is proposed to include eight transmembrane (TM) domains, which display the greatest degreee of conservation between PS-1 and PS-2. TM6 and TM7 are linked by a long acidic loop of relatively low conservation and variable length. The topology of the presenilins was determined using selective permeabilization with the pore-forming toxin streptolysin O, which does not disrupt the endoplasmic reticulum/Golgi membranes, followed by epitope-specific antibody tagging, which demonstrated that the N- and C-termini of the protein are orientated towards the cytoplasm [15]. This result is consistent with the absence of an obvious signal peptide at the N-terminus of the presenilins. The proposed TM structure of the presenilins (Figure 2) is based largely on work performed on the *Caenorhabditis elegans* protein sel-12 (see below), which displays a high degree of identity with PS-1 and PS-2. This study used a β-galactosidase hybrid protein approach, based on the observa-

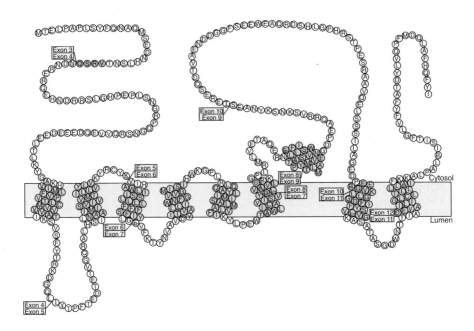

Figure 2. Structure of the PS-1 protein, showing exon boundaries and likely positions of the TM domains

The N- and C-termini are cytoplasmic. The membrane is believed, most usually, to be intra-cellular rather than at the cell surface. Residues within the eight proposed TM domains and the putative perimembranous domain are shaded blue. The VRSQ motif that is included or excluded depending on alternative splicing is shaded grey.

tion that β-galactosidase is active within the cytoplasm of cells but not in the extracytosolic compartment. Constructs containing the SEL-12 cDNA truncated after each of ten predicted TM domains (based on hydrophobicity analysis of sel-12) fused to *lacZ* were used to generate transgenic worms. The worms were then stained for β-galactosidase activity [16]. Only in constructs where the β-galactosidase was orientated into the cytosolic compartment was staining observed. This study demonstrated which putative TM domains actually spanned the membrane; to confirm the results, a synthetic TM domain was also added to each construct, which reversed the results of the original construct. As a result, eight of the nine putative TM domains were found to span the membrane (Figure 2).

Both PS-1 and PS-2 are proteolytically cleaved by an as yet unidentified protease(s) to generate two polypeptides [17–19]. In PS-1, the N-terminal fragment is 27–28 kDa and the C-terminal fragment is 17–18 kDa; the full-length protein is observed as an approx. 43 kDa species. The full-length PS-2 protein (53–55 kDa) is cleaved into a 35 kDa N-terminal fragment and a 20 kDa C-terminal fragment [20]. For both presenilins the predominant species that are observed in cultured mammalian cells and in the brain are the processed fragments. The major site of proteolytic cleavage in PS-1 has been localized to residue 292, corrresponding to exon 9, although cleavage at other adjacent

minor positions (291–299) is also observed, generating a heterogeneous population of N-terminal and C-terminal fragments [21]. The presenilins are also
thought to be degraded via a proteasome-mediated pathway after ubiquitination of the full-length protein, although the relationship of the two processing
and degradation pathways has not as yet been determined (see below) [22].

More than 40 mutations have been described in PS-1 that cause early-onset
AD, while only two have been described in PS-2. Mutations in PS-1 cause disease with an early (28–60 years) and consistent age of onset, while the age of
onset in PS-2 families is later and more variable (35–82 years) [7,24]. Only one
PS-1 mutation (Ile-143→Phe; in an English family) has been described with
possible incomplete penetrance [23]. At present there is no evidence that
apolipoprotein E (ApoE) genotype affects the age of onset in PS-1 families
[24], although there does appear to be a weak association between the rate of
disease progression and ApoE4 in the PS-2 Volga German families [25]. All
mutations in both PS-1 and PS-2 are missense mutations that affect residues
conserved between the two presenilins. The one exception is a mutation in
PS-1 that destroys the splice acceptor site for exon 9 (usually referred to as the
Δ9 mutation); this in turn results in the in-frame deletion of exon 9 from the
mutant transcripts and in an amino acid substitution (Ser-290→Cys) [26]. The
loss of exon 9 from the PS-1 mRNA also has the effect of removing the protease cleavage site from the protein, with the result that PS-1 Δ9 is not
processed in any of the mammalian cell lines into which it has been transfected
[18]. This mutation has been observed in families from England, Japan and
Australia [26–28]. The Australian kindred has an interesting phenotype in
which the presenting symptom is usually paraplegia [27]. Whether this is a
direct consequence of the nature of the PS-1 Δ9 mutation has yet to be determined. None of the other PS-1 or PS-2 mutations block presenilin processing
in all cell types (see [17]) in this manner, although it has been suggested that
the mutations lead to increased levels of the N- and C-terminal fragments by
partially blocking proteasome degradation of the presenilin protein [20]. No
nonsense or frameshift mutations in the presenilins have been observed, and
the Δ9 mutation preserves the open reading frame. Therefore it seems unlikely
that the mutations cause disease through a simple loss of function, since mutations that would be expected to block presenilin function completely have not
been found. A more likely explanation would appear to be that the mutations
cause a gain of dysfunction, or that the mutant protein blocks the function of
the wild-type protein through a dominant-negative mechanism.

Mutations have been identified at conserved residues throughout PS-1.
However, there are at least two distinct clusters of mutations. One of these is
at the N-terminal end of TM2 [9,29], while the other cluster extends from
TM6 into the N-terminal region of the long acidic loop between TM6 and
TM7 (exon 8) [30,31]. The proximity of the second cluster to the site of cleavage in the presenilin protein may be significant with regard to the effect of
these mutations on the function of PS-1. The mutations in TM2 in PS-1 are all

predicted to lie on one face of the proposed transmembrane α-helix; this pattern also extends to the residue (PS-1 Asn-135/PS-2 Asn-141) that is mutated in both presenilins [8,29]. It seems likely that the alignment of these mutations underlies their pathogenicity. Mutations in the two clusters give significantly different ages of onset when compared with each other: the TM2 cluster has a lower mean age of onset than the cluster around TM6 (mutations in the rest of the protein have an even higher mean age of onset). However, while multiple factors, not just related to the presenilin mutation, will affect the age of disease onset in a family, it appears likely that it is the nature of the amino acid substitution at a given site that has the more significant effect, rather than the position of the mutation in the protein. A clear example of this is at position Ile-143 in PS-1; the substitution Ile-143→Thr in a Belgian family gives an onset age of approx. 30 years, while the substitution Ile-143→Phe in an English family results in an onset age of approximately 55 years, and is also present in one elderly unaffected individual [23,30].

The prevalence of presenilin mutations in AD has been difficult to estimate, due to the effect of ascertainment bias in the collection of early-onset AD samples, which has clearly favoured large families with apparent autosomal dominant inheritance [31]. It is clear that initial suggestions that PS-1 mutations accounted for up to 70% of early-onset (below 65 years) AD cases were considerable overestimates. In a study of a Dutch early-onset AD series, the frequency of PS-1 mutations was 18% in families with autosomal dominant inheritance, 9% in familial AD with no clear inheritance pattern, and 6% in early-onset AD as a whole [32].

In our series, the occurrence of AD in all families with clear autosomal dominant inheritance has now been explained by mutations in either PS-1 (14 mutations; 14 families) or APP (three mutations; five families). The remaining early-onset families are too small to determine a clear mode of inheritance; however, significantly, these remaining families have excess levels of the ApoE4 allele [31,33]. Even with allowances for ascertainment bias, it appears certain that PS-1 mutations are more common than mutations in either APP or PS-2, which together probably account for less than 20% of the remaining autosomal dominant cases.

Effect of presenilin mutations on Aβ42(43)

Mutations in both PS-1 and PS-2 are associated with increased production of Aβ42(43), the amyloidogenic form of Aβ that is deposited selectively and early in AD [34]. This observation was initially made in samples of plasma and in the media from cultured fibroblasts from patients in families with autosomal dominant AD compared with control subjects [35]. A similar increase is not found in the level of Aβ40. Measurements of Aβ42(43) and Aβ40 were made in families with three different PS-1 mutations and in a Volga German family with the PS-2 mutation, using a sandwich ELISA assay that specifically detects the different forms of Aβ. This increase in Aβ42(43) was similar to that described in the same study in the plasma from members of families with mutations in residue 717 of APP, which also do not significantly affect the level of Aβ40. The 'Swedish' APP mutation increases the levels of both Aβ42(43) and Aβ40. Thus all the mutations, both in APP and in the presenilins, increase the level of Aβ42(43), suggesting a common pathological pathway by which these mutations cause AD [35,36]. In contrast, the majority (90%) of plasma samples from late-onset 'sporadic' AD cases did not show an increase in Aβ42(43), which suggests that the increase in the presenilin and APP families is a consequence of the mutations and not an indirect consequence of the disease state [35].

Subsequent studies have demonstrated increased production of Aβ42(43) in transfected cell lines and also in the brains of transgenic mice expressing mutant PS-1 and PS-2 cDNAs, consistent with the original observation in patient fibroblasts and plasma [37–40]. However, there is no clear link between the level of Aβ42(43) observed with each mutation and the severity of the associated phenotype ([38]; M. Hutton, J. Pérez-Tur and J. Hardy, unpublished work), as represented by the age of disease onset in families in which the mutations were identified. Many factors are probably responsible for this, including variations in clinical ascertainment and also the effect of different genetic backgrounds on age of onset. Interestingly the Δ9 mutation has consistently been found to give the highest Aβ42(43) levels ([38]; M. Hutton, J. Pérez-Tur and J. Hardy, unpublished work), when compared with PS-1 missense mutations, in transfected cell analysis. This is despite the fact that the age of onset in families bearing this mutation is not particularly early, and is another demonstration of the unusual nature of this mutation.

The mechanism by which the presenilin mutations alter the production of Aβ42(43) has yet to be determined; one possibility is that a direct interaction between presenilins and APP occurs that causes a subtle alteration in the cleavage of APP by single or multiple γ-secretase enzymes. Two observations are consistent with this hypothesis: first, the presenilins and APP are co-localized in the endoplasmic reticulum and early Golgi; secondly, it has been demonstrated that PS-2 and APP form stable complexes in transfected cells [41]. In this study, mammalian cells were first co-transfected with APP and PS-2. Immunoprecipitation of PS-2 revealed that a proportion of APP was associated with PS-2 immunocomplexes: similarly, precipitation of APP revealed associated PS-2 molecules. The interaction was non-covalent and was restricted to immature forms of APP, suggesting that the interaction occurred during the transit of APP through the endoplasmic reticulum, consistent with the subcellular localization of PS-2. The same report [41] also described a decrease in the secretion of APP in response to the co-expression of PS-2, suggesting that the presenilins are directly involved in the trafficking and processing of APP. However, the interaction between APP and PS-2 was not obviously affected by the Asn-141→Ile (Volga German) mutation, although a subtle effect that was undetectable in the immunoprecipitation paradigm could not be ruled out [41]. Thus the mechanism by which the mutations affect APP processing remains unclear. The presenilin mutations may directly influence γ-secretase cleavage of APP while complexed with PS-1 or PS-2, or the mutations may cause a subtle alteration in the trafficking of APP, such that a greater proportion enters a pathway that leads to the generation of Aβ42(43).

Early-onset autosomal dominant AD

Early-onset disease caused by mutations in APP and the presenilins is most probably related to an increase in the production of Aβ42(43), since all mutations in each of these genes lead to this effect. It seems unlikely that this increase is an epiphenomenon, as Aβ42(43) is the more amyloidogenic species of Aβ that is deposited selectively and early in AD [42]; in addition, the increased tendency of Aβ42(43) to form fibrils [43] also means that this species is potentially the more cytotoxic form of Aβ.

It is not clear whether the effects of the presenilin mutations on Aβ42(43) production reflect a specific effect of the presenilins on APP processing. The data derived from the analysis of the *Caenorhabditis elegans* homologues strongly suggest that the functions of the presenilins do not relate uniquely to APP, but rather that these proteins have a more general role, possibly in the intracellular trafficking of membrane proteins. This suggestion is consistent with the proposed involvement of the presenilins in apoptosis [44–46] and in embryonic development through the Notch pathway [47–53]. However, the observed interaction between PS-2 and APP suggests that the involvement of presenilins in APP processing, although not specific, is direct. If this is the case, the role of the presenilins in AD may merely reflect a minor alteration in presenilin function on protein processing whose only obvious consequence is an increase in Aβ42(43), which leads to disease after approximately 40 years.

Although all patients with early-onset AD caused by mutations in APP or the presenilins display increased production of Aβ42(43), only a small proportion (<10%) of typical AD cases show a similar increase [35]. These late-onset cases comprise >95% of all AD cases. It is not clear how the aetiology of these patients relates to the aetiology of the autosomal dominant cases. However, sporadic/late-onset AD has virtually identical pathology to that of AD caused by mutations in either the presenilins or APP [1], with both early and selective deposition of Aβ42(43) and the formation of both neuritic plaques and tangles. It therefore seems highly likely that similar pathways to pathogenesis are at work in both forms of the disease. One possibility is that proteins responsible for the 'clearance' of soluble Aβ or that accelerate deposition of Aβ are factors in the generation of disease in late-onset cases. Thus a minority of late-onset cases (<10%) display increased Aβ42(43) levels and presumably develop disease in a manner identical to that in individuals with APP or presenilin mutations, while the majority of late-onset cases (>90%) do not display increased Aβ42(43) levels but do undergo amyloidogenesis (despite apparently 'normal' Aβ levels) due to either poor clearance of Aβ or the presence of other factors that accelerate fibrillogenesis.

ApoE and other genetic risk factors for AD

Mutations in APP and the presenilins lead to AD in a simple, autosomal dominant fashion. That is, the presence of a single copy of the mutation leads, apparently inevitably, to disease. However, mutations in these genes account for a tiny proportion of cases of AD, and almost none of the typical, late-onset form of the disease.

The most common form of the disease occurs after 65 years of age, and a large effort has been expended to uncover the causes of this form of the disease. Epidemiological studies have suggested a number of risk factors for the disease, such as head trauma and level of education, but have also, most impressively, shown that the existence of a family history of dementia represents an increase in the risk of developing AD of 3–4-fold compared with the general population. This result indicates the possible existence of genetic factors influencing the appearance of the disease. The first genetic studies showed an association between familial late-onset AD and a gene located on chromosome 19, the ApoCII gene [54]. Subsequent to this, genetic linkage between this region of chromosome 19 and AD was reported [55]. Further biochemical experiments confirmed this observation and assigned the genetic effect to a different gene also located in that region of the chromosome, the ApoE gene [56]. The ApoCII gene is adjacent to the ApoE gene, and the original association was almost certainly due to linkage disequilibrium between the two loci.

ApoE is a polymorphic protein that has been extensively studied due to its role in cholesterol distribution between organs (for a review, see [57]). It is one of the major components of low-density lipoprotein (LDL). ApoE has three major isoforms, which differ in two amino acid positions. The most common isoform, E3, has a Cys residue at position 112 and an Arg at position 158. The two other variants contain either two Cys residues (E2) or two Arg residues (E4) at these positions. The different isoforms have different affinities for the receptor for LDL. The genetic data point towards an important increase in the risk of developing AD in those individuals bearing at least one copy of the ε4 allele. In addition, there is dose effect for this allele in two major aspects. Individuals homozygous for the allele usually present the onset of the disease at a younger age than heterozygous individuals; the latter, in turn, have a younger age of onset than cases without a copy of this allele [58]. In general, it seems that each ε4 allele shifts the risk curve for developing AD forward by 5 years, and that each ε2 allele shifts the risk curve backwards by a similar amount, relative to ε3 homozygotes [58,59]. In addition, the extension of Aβ deposition in the brain parenchyma can also be related to the number of copies of the ε4 allele. The same relationship was observed, i.e. cases with two copies of the ε4 allele had more neuritic deposits in their brains than cases with one copy, which in turn had more than cases with no copies of this allele [60]. Finally, several studies have suggested the possibility that the ε2 allele could act in an opposite manner to the ε4 allele. The presence of this allele seems to reduce the risk of the disease [59,61,62], although this effect is not as strong as the one observed for the ε4 allele and is not always observed in different populations.

Although its relationship to AD is not obvious, isoform-specific effects between ApoE and the Aβ peptide or tau protein have both been reported, suggesting direct effects of ApoE in the disease [56,63]. Other isoform-specific events have also been reported, notably the differential effects of the E3 and E4 isoforms in promoting neuronal growth in cell culture [64].

However, *a priori* perhaps it is most likely that ApoE's role in AD relates directly to its well-established role in lipid metabolism. ApoE is expressed largely by astrocytes, mainly as a response to an insult, in order to recycle the cholesterol in provenance of the plasma membrane. This cholesterol will be re-used afterwards in the regeneration process. This, together with the fact that binding to the receptor is not equivalent for each isoform, suggests that the effect of the ε4 allele can be explained as a reduced ability of the brain to regenerate after an insult. Thus it may well be that ApoE's role in AD relates not so much to the disease process itself, but rather to a more general role in repairing synaptic damage, and individuals who have ε4 alleles are less well able to respond appropriately to the process [65].

Are there other Alzheimer genes, and what are they likely to be?

Within families with early-onset autosomal dominant AD in our series, mutations have now been identified in all but two, and in these two there is clear linkage to the PS-1 locus, suggesting that there are more complex mutations at this site that we have yet to identify. Thus the vast majority of cases, if not all, of early-onset autosomal dominant AD have been explained. However, there are clearly many cases of familial disease, with both early and late onset, in which the genetic contribution to the disease is not yet worked out. Several other genes have been postulated to be involved in the disease, but none of the other reports have yet survived the test of replication. Other genetic loci have been proposed, most notably one to a locus on chromosome 12, but this has not yet been reported in the scientific literature, nor has it yet been replicated, and so must remain uncertain at this point.

Based on what has been discussed above, there are two clear types of gene that may be involved in modification or genetic risk for AD: those whose products modulate Aβ production, and those whose products are involved in cellular repair and response to damage. However, it has to be admitted that the other attractive feature of molecular genetics as a way to study AD has been that each new gene has been a surprise and has opened a new area of research; it is only now that we are beginning to understand the links between them.

Summary

- *Mutations in any one of three genes can cause autosomal dominant, early-onset Alzheimer's disease: these genes are the amyloid precursor protein (APP) gene on chromosome 21, the presenilin-1 (PS-1) gene on chromosome 14 and the presenilin-2 (PS-2) gene on chromosome 1.*
- *Pathogenic mutations at all these loci cause mismetabolism of APP such that more of the peptide Aβ42 is produced.*
- *This peptide is deposited in the plaques in the brains of Alzheimer's patients. These facts have led to the dominant hypothesis for the disease process: the 'amyloid cascade hypothesis', which proposes that overproduction or failure to clear the peptide Aβ42 is always central to the disease.*
- *Genetic variability at the apoliprotein E locus is a major determinant of late onset Alzheimer's disease.*
- *The mechanism by which apoliprotein E is involved in the pathogenesis of Alzheimer's disease is not yet known.*
- *There are likely to be other genetic factors which impinge on Alzheimer's disease.*

References

1. St. George Hyslop, P., Haines, J., Farrer, L. et al. (1990) Genetic linkage studies suggest that Alzheimer's disease is not a single homogenous disorder. *Nature (London)* **347**, 194–197
2. Goate, A.M., Chartier-Harlin, M.C. & Mullan, M.J. (1991) Segregation of a missense mutation in the amyloid precursor protein gene with familial Alzheimer's disease. *Nature (London)* **349**, 704–706
3. Selkoe, D.J. (1996) Amyloid ß-protein and the genetics of Alzheimer's disease. *J. Biol. Chem.* **271**, 18295–18298
4. Glenner, G.G. & Wong, C.W. (1984) Alzheimer's disease: initial report of the purification and characterization of a novel cerebrovascular amyloid protein. *Biochem. Biophys. Res. Commun.* **120**, 885–890
5. Masters, C.L., Simms, G., Weunman, N.A. et al. (1985) Amyloid plaque core protein in Alzheimer's disease and Down syndrome. *Proc. Natl. Acad. Sci. U.S.A.* **82**, 4245–4249
6. Sherrington, R., Rogaev, E.I., Liang, Y. et al. (1995) Cloning of a gene bearing missense mutations in early-onset familial Alzheimer's disease. *Nature (London)* **375**, 754–760
7. Cruts, M., Backhovens, H., Wang, S.Y. et al. (1995) Molecular genetics analysis of familial early-onset Alzheimer's disease linked to chromosome 14q24.3. *Hum. Mol. Genet.* **12**, 2363–2371
8. Levy-Lahad, E., Wasco. W., Poorkaj, P. et al. (1995) Candidate gene for the chromosome 1 familial Alzheimer's disease locus. *Science* **269**, 973–977
9. Clarke, R.F., Hutton, M. et al. (1995) The structure of the presenilin-1 (S182) gene and identification of six novel mutations in early onset AD families. *Nature Genet.* **11**, 219–222

10. Rogaev, E.I., Sherrington, R., Rogaeva, E.A. et al. (1995) Familial Alzheimer's disease in kindreds with missense mutations in a gene on chromosome 1 related to the Alzheimer's disease type 3 gene. *Nature (London)* **376**, 775–778

11. Prihar, G., Fuldner, R.A., Perez-Tur, J. et al. (1996) Structure and alternative splicing of the presenilin-2 gene. *NeuroReport* **7**, 1680–1684

12. Levy-Lahad, E., Poorkaj, P., Wang, K. et al. (1996) Genomic structure and expression of STM2, the chromosome 1 familial Alzheimer's disease gene. *Genomics* **34**, 198–204

13. Rogaev, E.I., Sherrington, R., Wu, C. et al. (1997) Analysis of the 5' sequence, genomic structure and alternative splicing of the presenilin 1 gene associated with early onset Alzheimer's disease. *Genomics*, **40**, 415–424

14. Perez-Tur, J., Croxton, R., Wright, K. et al. (1996) A further presenilin 1 mutation in the exon 8 cluster in familial Alzheimer's disease. *Neurodegeneration* **5**, 207–212

15. Doan, A., Thinakaran, G., Borchelt, D.R. et al. (1996) Protein topology of presenilin 1. *Neuron* **17**, 1023–1030

16. Li, X. & Greenwald, I. (1996) Membrane topology of the *C. elegans* SEL-12 presenilin. *Neuron* **17**, 1015–1021

17. Mercken, M., Takahashi, H., Honda, T. et al. (1996) Characterization of human presenilin 1 using N-terminal specific monoclonal antibodies: evidence that Alzheimer mutations affect proteolytic processing. *FEBS Lett.* **389**, 297–303

18. Thinakaran, G., Borchelt, D.R., Lee, M.K. et al. (1996) Endoproteolysis of presenilin 1 and accumulation of processed derivatives *in vivo*. *Neuron* **17**, 181–190

19. Ward, R.V., Davis, J.B., Gray, C.W. et al. (1996) Presenilin 1 is processed into two major cleavage products in neuronal cell lines. *Neurodegeneration* **5**, 293–298

20. Tanzi, R.E., Kovacs, D.M., Tae-Wan, K. et al. (1996) The presenilin genes and their role in early-onset familial Alzheimer's disease. *Alzheimer's Dis. Rev.* **1**, 90–98

21. Podlisny, M., Citron, M., Amarante, P. et al. (1997) Presenilin proteins undergo heterogeneous endoproteolysis between Thr291 and Ala299 and occur as stable N- and C-terminal fragments in normal and AD brain tissue. *Neurobiol. Dis.* **3**, 325–337

22. Kim, T.W., Pettingell, W.H., Moir R.D., Wasco, W. & Tanzi, R.E. (1997) The presenilin genes and their role in early onset familial Alzheimer's disease. *J. Biol. Chem.* **272**, 11006–11010

23. Rossor, M.N., Fox, N.C., Beck, J., Campbell, T.C. & Collinge, J. (1996) Incomplete penetrance of familial Alzheimer's disease in a pedigree with a novel presenilin 1 gene mutation. *Lancet* **347**, 1560

24. Van Broeckhoven, C., Backhovens, H., Cruts, M. et al. (1994) ApoE genotype does not modulate age of onset in families with chromosome 14 encoded Alzheimer's disease. *Neurosci. Lett.* **169**, 179–180

25. Bird, T.D., Levy-Lahad, E., Poorkai, P. et al. (1997) Wide range in age of onset for chromosome 1-related familial Alzheimer's disease. *Ann. Neurol.* **40**, 932–936

26. Perez-Tur, J., Froelich, S., Prihar, G. et al. (1995) A mutation in Alzheimer's disease destroying a splice acceptor site in the presenilin 1 gene. *NeuroReport* **7**, 204–207

27. Kwok, J.B.J., Taddel, K., Hallupp, M. et al. (1997) Two novel (M233T and R278T) presenilin 1 mutations in early onset Alzheimer's disease and preliminary evidence for association of presenilin 1 mutations with a novel phenotype. *NeuroReport*, in the press

28. Sherrington, R., Froelich, S., Sorbi, S. et al. (1996) Alzheimer's disease associated with mutations in presenilin 2 is rare and variably penetrant. *Hum. Mol. Genet.* **5**, 985–988

29. Crook, R., Ellis, R., Shanks, M. et al. (1997) Early onset Alzheimer's disease with a presenilin 1 mutation at the site corresponding to the Volga German presenilin 2 mutation. *Ann. Neurol.*, **42**, 124–128

30. Cruts, M., Hendriks, L. & Van Broeckhoven, C. (1996) The presenilin genes: a new gene family involved in Alzheimer disease pathology. *Hum. Mol. Genet.* **5**, 1449–1455

31. Hutton, M., Busfield, F., Wragg, M. et al. (1996) Complete analysis of the presenilin 1 gene in families early onset Alzheimer's disease. *NeuroReport* **7**, 801–805

32. Cruts, M., van Duijn, C.M., Backhovens, H. et al. (1998) Estimation of the genetic contribution of presenilin-1 and -2 mutations in a population-based study of presenile Alzheimer disease. *Hum. Mol. Genet.* **7**, 43–51

33. Houlden, H., Crook, R., Backhovens, H. et al. (1998) ApoE genotype is a risk factor in non-presenilin early onset Alzheimer's disease families. *Neuropsychiatr. Genet.*, **81**, 117–121

34. Iwatsubo, T., Odaka, A., Suzuki, N. et al. (1994) Visualization of Aβ42(43) and Aβ40 in senile plaques with end-specific Aβ monoclonals: evidence that an initially deposited species is Aβ42(43) *Neuron* **13**, 45–53

35. Scheuner, D., Eckman, C., Jensen, M. et al. (1996) Secreted amyloid beta-protein similar to that in the senile plaques of Alzheimer's disease is increased *in vivo* by the presenilin 1 and 2 and APP mutations linked to familial Alzheimer's disease. *Nat. Med.* **2**, 864–870

36. Hardy, J. (1997) Amyloid, the presenilins and Alzheimer's disease. *Trends Neurosci.* **20**, 154–159

37. Duff, K., Eckman, C., Zehr, C. et al. (1996) Increased amyloid-β42(43) in brains of mice expressing mutant presenilin 1. *Nature (London)* **383**, 710–713

38. Citron, M., Westaway, D., Xia, W. et al. (1997) Mutant presenilins of Alzheimer's disease increase production of 42-residue amyloid β-protein in both transfected cells and transgenic mice. *Nat. Med.* **3**, 67–72

39. Tomita, T., Maruyama, K., Saido, T.C. et al. (1997) The presenilin 2 mutation (N141I) linked to familial Alzheimer disease (Volga German families) increases the secretion of amyloid β protein ending at the 42nd (or 43rd) residue. *Proc. Natl. Acad. Sci. U.S.A.* **94**, 2025–2030

40. Borcheldt, D.R., Thinakaran, G., Eckman, C.B., Lee, M.K., Davenport, F., Ratovitsky, T., Prada, C.M., Kim,. G., Seekins, S., Yager, D., et al. (1996) Familial Alzheimer's disease-linked presenilin 1 variants elevate A β1-42/1-40 ratio in vitro and in vivo. *Neuron* **17(5)** 1005–1013

41. Weidemann, A., Paliga, K., Durrwang, U. et al. (1997) Formation of stable complexes between two Alzheimer's disease gene products: presenilin-2 and β-amyloid precursor protein. *Nature Med.* **3**, 328–332

42. Mann, D.M., Iwatsubo, T., Cairns, N.J. et al. (1996) Amyloid-β protein (Aβ) deposition in chromosome 14-linked Alzheimer's disease: predominance of Aβ42(43). *Ann. Neurol.* **40**, 149–156

43. Jarrett, J.T., Berger, E.P. & Lansbury, P.T. (1993) The carboxy terminus of β-amyloid protein is critical for the seeding of amyloid formation: implications for the pathogenesis of Alzheimer's disease. *Biochemistry* **32**, 4693–4697

44. Vito, P., Wolozin, B., Ganjei, J.K. et al. (1996) Requirement of the familial Alzheimer's disease gene PS-2 for apoptosis. Opposing effect of ALG-3. *J. Biol. Chem.* **271**, 31025–31028

45. Vito, P., Lacana, E. & D'Adamio, L. (1996) Interfering with apoptosis: Ca²⁺-binding protein ALG-2 and Alzheimer's disease gene ALG-3. *Science* **271**, 521–525

46. Wolozin, B., Iwasaki, K., Vito, P. et al. (1996) Participation of presenilin 2 in apoptosis: enhanced basal activity conferred by an Alzheimer mutation. *Science* **274**, 1710–1713

47. Levitan, D. & Greenwald, I. (1995) Facilitation of *lin-12*-mediated signalling by *Sel-12*, a *Caenorhabditis elegans* S182 Alzheimer's disease gene. *Nature (London)* **377**, 351–354

48. L'Hernault, S.W. & Arduengo, P.M. (1992) Mutation of a putative sperm membrane protein in *Caenorhabditis elegans* prevents sperm differentiation but not its associated meiotic divisions. *J. Cell Biol.* **119**, 55–68

49. Wong, P.C., Zheng, H., Chen, H. et al. (1997) Presenilin 1 is required for Notch1 and Dll1 expression in the paraxial mesoderm. *Nature (London)* **387**, 288–292

50. Hrabe de Angelis, M., McIntyre, II, J. & Gossler, A. (1997) Maintenance of somite borders in mice requires the Delta homologue Dll1. *Nature (London)* **386**, 717–728

51. Shen, J., Bronson, R.T., Chen, D.F. et al. (1997) Skeletal and CNS defects in presenilin-1-deficient mice. *Cell* **89**, 629–639

52. Levitan, D., Doyle, T.G., Brousseau, D. et al. (1996) Assessment of normal and mutant human presenilin function in *Caenorhabditis elegans*. *Proc. Natl. Acad. Sci. U.S.A.* **93**, 14940–14944

M. Hutton et al.

131

53. Baumeister, R., Leimer, U., Zweckbronner, I. et al. (1997) Proteolytic cleavage of the Alzheimer's disease associated presenilin 1 is not required for its function in *Caenorhabditis elegans* notch signalling. *Genes Function*, **1**, 139–147

54. Schellenberg, G.D., Deeb, S.S., Boehnke, M. et al. (1987) Association of an apolipoprotein CII allele with familial dementia of the Alzheimer type. *J. Neurogenet.* **4**, 97–108

55. Pericak-Vance, M.A., Bebout, J.L., Gaskell, Jr., P.C. et al. (1991) Linkage studies in familial Alzheimer disease: evidence for chromosome 19 linkage. *Am. J. Hum. Genet.* **48**, 1034–1050

56. Strittmatter, W.J., Weisgraber, K.H., Huang, D.Y. et al. (1993) Binding of human apolipoprotein E to synthetic amyloid beta peptide: isoform-specific effects and implications for late-onset Alzheimer disease. *Proc. Natl Acad. Sci. U.S.A.* **90**, 8098–8102

57. Mahley, R.W. (1988) Apolipoprotein E: cholesterol transport protein with expanding role in cell biology. *Science* **240**, 622–630

58. Corder, E.H., Saunders, A.M., Strittmatter, W.J. et al. (1993) Gene dose of apolipoprotein E type 4 allele and the risk of Alzheimer's disease in late onset families. *Science* **261**, 921–923

59. Houlden, H., Collinge, J., Kennedy, A. et al. (1993) ApoE genotype and Alzheimer's disease. *Lancet* **342**, 737–738

60. Schmechel, D.E., Saunders, A.M., Strittmatter, W.J. et al. (1993) Increased amyloid beta-peptide deposition in cerebral cortex as a consequence of apolipoprotein E genotype in late-onset Alzheimer disease. *Proc. Natl. Acad. Sci. U.S.A.* **90**, 9649–9653

61. Chartier-Harlin, M.C., Parfitt, M., Legrain, S. et al. (1994) Apolipoprotein E, e4 allele as a major risk factor for sporadic early- and late-onset forms of Alzheimer's disease: analysis of the 19q13.2 chromosomal region. *Hum. Mol. Genet* **3**, 569–574

62. Corder, E.H., Saunders, A.M., Risch, N.J., et al. (1994) Protective effect of apolipoprotein E type 2 allele for late onset Alzheimer's disease. *Nat. Genet.* **7**, 180–184

63. Strittmatter, W.J., Saunders, A.M., Goedert, M. et al. (1994) Isoform-specific interactions of apolipoprotein E with microtubule-associated protein tau: implications for Alzheimer's disease. *Proc. Natl. Acad. Sci. U.S.A.* **91**, 11183–11186

64. Nathan, B.P., Bellosta, S., Sanan, D.A. et al. (1994) Differential effects of apolipoproteins E3 and E4 on neuronal growth in vitro. *Science* **264**, 850–852

65. Poirier, J., Minnich, A. & Davignon, J. (1995) Apolipoprotein E, synaptic plasticity and Alzheimer's disease. *Ann. Med.* **27**, 663–670

11

Use of brain grafts to study the pathogenesis of prion diseases

Adriano Aguzzi[1], Michael A. Klein, Christine Musahl, Alex J. Raeber, Thomas Blättler, Ivan Hegyi, Rico Frigg and Sebastian Brandner

Department of Pathology, Institute of Neuropathology, University of Zürich, CH-8091 Zürich, Switzerland

Introduction

Prion diseases, such as scrapie, have long been thought to affect mainly sheep and some exotic animals, such as elk and mink. The epidemiology of human prion diseases is confined to Creutzfeldt–Jakob disease and related ailments, which are exceedingly rare, and to Kuru, which is unlikely to represent a danger to humans unless one indulges in cannibalism. However, a new variant of Creutzfeldt–Jakob disease (nvCJD) is now occurring in humans, which is thought to result from the ingestion of products contaminated with bovine spongiform encephalopathy (BSE). At the time of writing, only two dozen individuals have succumbed to nvCJD, but because many persons may have been exposed to the BSE agent, a thorough understanding of transmissible spongiform encephalopathies is clearly needed [1,2].

Two as yet unresolved questions continue to occupy the attention of researchers in the field. The first relates to the actual structure of the infectious agent, which has been named a prion and which replicates in the central nervous system (CNS) and in some other tissues of infected animals and humans.

[1]*To whom correspondence should be addressed.*

We have dealt with these issues in several recent review articles [1], and therefore will not focus on this discussion here.

The second question is no less intriguing: by which mechanisms can prions bring about damage to the CNS? Is the damage related to the actual replication of the prion? Or is the spongiform encephalopathy a result of the accumulation of toxic metabolites within, or around, neurons? One prime candidate for the toxicity hypothesis is certainly PrP^{Sc}, the pathologically changed isoform of the normal prion protein PrP^C. But is PrP^{Sc} toxic only if it is generated within cells, or can it damage nerve cells when acting from without? In addition, since prions appear to replicate in the organs of the lymphoreticular system, such as the spleen, lymph nodes and Peyer's plaques of the intestine, why are immune deficiencies or damage to these organs not encountered after infection with prions? In other words, is susceptibility to prion toxicity a unique property of neural tissue, or is it rather the result of the much higher levels of PrP^{Sc} accumulation seen in the brains, as compared with the lymphoreticular organs, of terminally sick scrapie mice?

It may be very difficult to devise suitable systems to address these issues experimentally. However, the recent generation of transgenic and knock-out mice for the *Prnp* gene has opened up new, promising avenues of investigation. We have taken advantage of various strains of transgenic mice, and have asked whether neurografting technology could be used to address the question of prion neurotoxicity.

Neural grafting (described in the next section) has often been used to address questions related to developmental neurobiology. Several studies have investigated the establishment of neuronal organization within grafts and interactions with the host. Recent grafting studies have been aimed at questions related to neural plasticity; for example, whether and to what extent undifferentiated progenitor cells could integrate and participate in the formation of the host CNS [3].

In the field of neurodegenerative disorders, grafting studies have been aimed mainly at reconstituting certain pathways or particular functions after surgical or toxic lesions to selected functional systems. In these models, an artificial lesion leads to the degeneration of specific neuronal systems. Grafting of neural tissue or genetically engineered cells aims at the functional repair of induced lesions [4]. Many such experiments have been carried out in the rat system, which is well suited for developmental studies and allows stereotaxic surgical interventions (where specific anatomical regions can be targeted following calculation of the co-ordinates) to be carried out with appropriate accuracy.

In this overview, we describe the biological properties of neuroepithelial grafts, such as tissue growth, proliferation and differentiation. Special emphasis is placed on the development of the blood–brain barrier (BBB) after grafting. We then describe how embryonic telencephalic grafting has been applied to the study of scrapie pathogenesis.

Biological characteristics of mouse neuroectodermal grafts

Transgenic techniques facilitate the study of the role of single molecules in development, and in functional and pathological processes. The generation of knock-out mice by the targeted deletion of genes of interest has expanded our understanding of the molecular mechanisms of neural development and the pathogenesis of CNS diseases. Transgenic and knock-out mice have provided valuable models for neurodegenerative diseases [5–7]. Others, however, show early postnatal [8–10] or even embryonic [11,12] lethal phenotypes which can be difficult to interpret. Lethal phenotypes provide strong evidence for a crucial role for the respective gene products during development, and hint at an important role for these factors in the determination of cell fates during differentiation [12–14,14a], but they do not allow us to study the roles that these factors play in secondary pathological processes such as neurodegeneration. In an effort to overcome this problem, we have employed transplantation approaches using neural tissue derived from such mouse embryos. Using grafting techniques, it has been possible to study the neural tissue of mice with premature lethal genotypes at time points that exceed by far the life span of the mutant mice [14,14a,15].

For the grafting procedure, neuroectodermal tissue is harvested from a mouse and implanted stereotaxically into the caudoputamen of recipient mice. Such neurografts contain neurons with myelinated processes and a dense synaptic network, glia (astrocytes, oligodendrocytes and microglia) and blood vessels 4 weeks after grafting (projected age of grafted tissue is approximately postnatal day 20) [14,14a,15]. Taken together, these findings indicate that embryonic neuroepithelial tissue grafted into an adult host brain follows a programme of maturation and differentiation with a time course similar to that occurring *in vivo* [14,14a].

Blood–brain barrier and brain grafts

The BBB maintains the homoeostatic environment in the brain by preventing blood-borne compounds gaining free entry into the CNS parenchyma. The barrier is formed by tight junctions in the vascular endothelia, which are probably induced by astrocytes. A number of pathological CNS processes, such as inflammation, demyelination, tumour growth or degeneration, can induce breakdown of the BBB. In turn, BBB leakage might induce CNS dysfunction caused by blood-borne neurotoxic compounds that are normally excluded from the brain parenchyma [16].

The post-transplantation status of the BBB in rodents is still controversial. An early, yet most valuable, study suggested that the type of donor tissue determines the characteristics and BBB properties of vessels supplying the graft [17]. According to this hypothesis, neural grafts induce BBB properties in the supplying blood vessels. In fact, several authors have described complete BBB reconstitution after neural grafting to the CNS. Some even found no

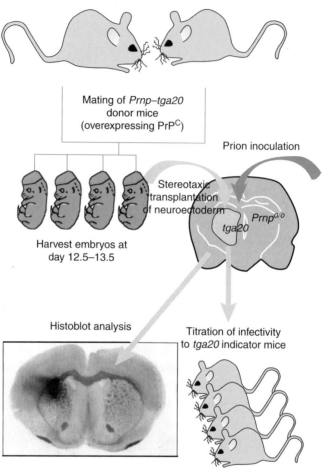

Figure 1. Use of the neurografting technique for the study of mouse scrapie
The neuroectodermal anlage from a PrP^C-overexpressing mouse (designated *tga20*) was implanted into the brain of a PrP-deficient knock-out mouse, and prions were administered intracerebrally. After an incubation time of at least 10 weeks, the brains were assayed for the
presence of proteinase-resistant PrP and for infectivity. Titration for the determination of
infectivity is performed by scratching graft tissue from cryosections, homogenization and intracerebral injection into PrP-overexpressing *tga20* mice. The presence of proteinase-resistant PrP
is detected by immunohistochemistry of adjacent sections.

residual BBB leakage as early as 1 week after grafting [18]. We carried out
studies in the model described using four independent marker molecules to
detect damage to the BBB. The results obtained with various techniques were
consistent. Both magnetic resonance imaging (MRI) and histological analysis
indicated that the BBB was reconstituted in 67% of all grafts after 3 weeks, and
in 90% of the grafts 7 weeks after grafting. These findings indicate that the
grafting procedure usually does not induce permanent BBB leakage that might
expose the grafted tissue to a non-physiological environment, and suggest that
the genotype of the grafted tissue determines the BBB properties of the graft.

Neurografts in prion research

We decided to apply the grafting technique to the study of scrapie. $Prnp^{o/o}$ mice, which are devoid of PrP^C, are resistant to scrapie and do not propagate prions [5,19]. Because these mice show normal development and behaviour [20], it has been argued that scrapie pathology may come about because PrP^{Sc} deposition is neurotoxic [21], rather than due to the depletion of cellular PrP^C. On the other hand, the acute depletion of PrP^C might be much more deleterious than its absence throughout development, since the organism may then not have time to activate compensatory mechanisms.

One way to address the question of neurotoxicity was to expose brain tissue of $Prnp^{o/o}$ mice to a continuous source of PrP^{Sc}. We therefore grafted neural tissue overexpressing PrP into the brains of PrP-deficient mice (Figure 1). After intracerebral inoculation of scrapie prions, the grafts accumulated high levels of PrP^{Sc} and infectivity (Table 1), and developed severe histopathological changes characteristic of scrapie. Substantial amounts of graft-derived PrP^{Sc}

Table 1. Determination of scrapie prion infectivity in grafts and in regions of the host brain adjacent to, or distant from, the graft

Source of infectivity	Time after inoculation (days)	Transmission	Incubation period in recipients (days)
(a) Standard prion inoculum (with log dilutions of inoculum)			
RML, 10^{-1}	–	4/4	58, 59, 62, 67
RML, 10^{-3}	–	2/2	65, 65
RML, 10^{-5}	–	4/4	83, 84, 84, 95
RML, 10^{-7}	–	1/3	109, >217, >217
RML, 10^{-9}	–	0/4	>217, >217, >217, >217
RML, 10^{-11}	–	0/3	>217, >217, >217
(b) Mouse tissue			
Graft region	245	2/2	75, 75
Contralateral frontal cortex	336	0/2	>170, >170
Contralateral frontal cortex	350	0/2	>170, >170
Contralateral frontal cortex	428	0/2	>170, >170
Contralateral frontal cortex	454	0/2	>170, >170
Contralateral parietal cortex	285	2/2	103, 121
Cerebellum	285	0/3	>170, >170, >170
Spleen	285	0/1	>170

The presence of prion infectivity is correlated with the presence of histologically detectable PrP deposits. Abbreviation: RML, Rocky Mountain laboratory.

migrated into the host brains ([22,22a,23]; see also Figure 1). However, even 16 months after transplantation and infection with prions, no pathological changes were detected in the PrP-deficient recipient tissue, even in the immediate vicinity of the grafts. Such results suggest that PrP^{Sc} is inherently non-toxic and that PrP^{Sc} plaques found in spongiform encephalopathies may be an epiphenomenon rather than a cause of neuronal damage. It is conceivable that PrP^{Sc} is only toxic when it is formed and accumulated within the cell, but not when presented from outside. Finally, it may be that PrP^{Sc} is pathogenic when presented from without, but only to cells expressing PrP^{C}, either because it initiates conversion of PrP^{C} into PrP^{Sc} at the cell surface and/or because it is internalized in association with PrP^{C}, which is endocytosed efficiently [24].

The fact that a scrapie-infected host animal harbouring a neuroectodermal graft does not develop clinical signs of scrapie allowed investigation of the pathological chain of events leading to neurodegeneration, and determination of the end point of the disease. Based on morphological criteria, it was found that, with increasing incubation periods, grafts underwent progressive astrogliosis and spongiosis, which were accompanied by loss of neuronal processes within the grafts and subsequent destruction of the neuropil (see below).

Spread of prions in the CNS

Intracerebral inoculation of tissue homogenate into suitable recipients is the most effective method for the transmission of spongiform encephalopathies. However, spongiform encephalopathies have also been transmitted by feeding, as well as by intravenous, intraperitoneal and intramuscular injection. Prion diseases can also be initiated from the eye by conjunctival instillation, corneal grafts and intraocular injection [25]. The latter method has proved particularly useful in studying the neural spread of the agent, since the retina is a part of the CNS, and intraocular injection leads to the progressive appearance of pathological changes typical of scrapie along the optic tract, the lateral geniculate nucleus and the superior colliculus; eventually the CNS is colonized, leading to generalized spongiform encephalopathy (Figures 2b–2d).

To determine whether the spread of prions along the optic pathway is dependent on the presence of PrP^{C}, we again used grafting technology. This time, the neural grafts served as indicators of the transport of prion infectivity from the retina to the CNS. After intraocular inoculation, *Prnp$^{o/o}$* mice grafted with *Prnp*-overexpressing tissue were allowed to survive for up to 1.5 years. Even at this time, none of the mice showed any sign of spongiform encephalopathy in their grafts. Therefore PrP^{C} appears to be necessary for the spread of prions from the eye to the rest of the CNS.

Engraftment of *Prnp$^{o/o}$* mice with PrP^{C}-producing tissue may conceivably lead to an immune response to PrP [26], and possibly to neutralization of infectivity. Indeed, sera from grafted mice had significant anti-PrP antibody

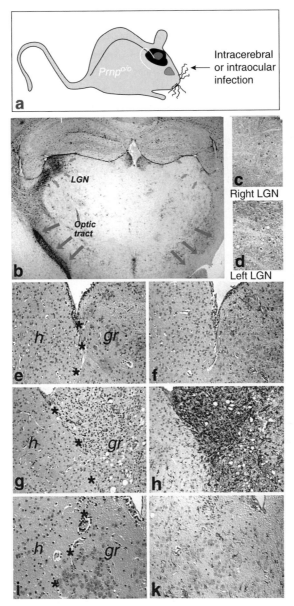

Figure 2. Tracing the spread of prions in the nervous system of PrP-deficient hosts
(a) Schematic drawing of the transplantation procedure. (b)–(d) Coronal sections of the thalamus of a PrP-overexpressing *tga20* mouse 78 days after inoculation of prions into the right eye. At the time of analysis, the animal showed clinical symptoms of scrapie. (b) Pronounced gliosis in the visual pathway [optic tract and lateral geniculate nucleus (LGN)] visualized by immunocytochemistry for glial fibrillary acidic protein (GFAP). (c, d) Asymmetric neurodegeneration of the LGN visualized by synaptophysin immunostaining. Coarse granular deposits and patchy staining reflect significant synaptic loss in the affected left LGN, while the right, unaffected LGN displays the fine granular synaptic staining typical of normal neural tissue. Because scrapie infection starts in the optic tract and is followed by generalized disease in the CNS, the LGN and superior colliculus show a more prominent astrocytic reaction and a more severe loss of neuronal processes than other regions of the brain, e.g. the hippocampus. (e)–(k) Paraffin sections taken from grafted *Prnp^{o/o}* mice 232 days after mock treatment (no infection) (e, f), 230 days after intracerebral inoculation of prions (g, h) and 217 days after intraocular inoculation of prions (i, k). Abbreviations: h, host; gr, graft. Magnification ×40.

titres. To test whether grafts would develop scrapie if infectivity were administered before the establishment of an immune response, we inoculated mice 24 h after grafting. Again, no disease was detected in the grafts of two mice inoculated intraocularly.

In order to rule out definitively the possibility that prion transport had been disabled by a neutralizing immune response, we repeated the experiments in mice largely tolerant of PrP. For this, we introduced a transgene into $Prnp^{o/o}$ mice that led to the overexpression of PrP on T-lymphocytes. These mice were still resistant to scrapie and, when engrafted with PrP^C-overexpressing neuroectoderm, did not develop antibodies to PrP, presumably due to the clonal deletion of PrP-immunoreactive lymphocytes. As before, intraocular inoculation with prions did not provoke scrapie in the graft, supporting the conclusion that lack of PrP^C, rather than an immune response to PrP, prevents spread.

Pathological findings characteristic of scrapie, and replication of infectivity after intraocular injection, occur along the anatomical structures of the visual system [25] and spread to trans-synaptic structures (such as the contralateral superior colliculus, lateral geniculate nucleus and visual cortex; see Figure 2b). This has been taken as evidence for axonal transport of the agent. However, although PrP^C seems to travel with fast axonal transport [27], the very slow kinetics of disease development caused by prions suggests a different mode of spread for infectious prions. Since intraocular inoculation failed to infect grafts even in the absence of an immune response to PrP (Figures 2i and 2k), PrP^C appears to be necessary for the spread of prions along the retinal projections and within the intact CNS. The prion itself seems, therefore, to be surprisingly sessile.

Since prion infectivity is consistently detectable in the spleen earlier than in the brain, even after intracerebral inoculation [28], it could be argued that prion replication in lymphoreticular organs may be involved in the neuroinvasiveness of intraocularly administered prions. Enucleation as late as 7 days following intraocular inoculation resulted in scrapie, but prevented targeting to the visual system [29]. This suggests that systemic infection and secondary neuroinvasion can bypass the neural spread of prions if the visual pathway is interrupted before prions can colonize the brain via the optical pathway. Therefore the lack of graft infection following intraocular prion inoculation suggests that the absence of extracerebral PrP^C impairs prion spread from extracerebral sites to the CNS, in addition to blocking neural spread.

Thus the intracerebral spread of prions is based on a PrP^C-paved chain of cells, perhaps because such cells are capable of supporting prion replication [22,22a]. When such a chain is interrupted by interposed cells that lack PrP^C, as in the case described here, no propagation of prions to the target tissue can occur. Perhaps prions require PrP^C for propagation across synapses: PrP^C is present in the synaptic region and certain synaptic properties are altered in $Prnp^{o/o}$ mice [30]. Perhaps transport of prions within (or on the surface of) neuronal processes is PrP^C-dependent. Within the framework of the protein-only hypothesis, these findings may be accommodated by a 'domino-effect' model, whereby the spread of scrapie prions in the CNS occurs continuously through the conversion of PrP^C by adjacent PrP^{Sc} [31].

Cells in the CNS that are affected by spongiform encephalopathies

Transmissible spongiform encephalopathies show typical histopathological features in the CNS, such as vacuolization and reactive astrogliosis. Concurrent neuronal loss is frequently observed [32], but in some instances (e.g. different strains of the agent) this feature is not obvious [33] and is detectable only after detailed morphometric analysis or in specific experimental systems [34].

Mice expressing increased amounts of PrP^C in their brains develop scrapie with a shortened incubation time after intracerebral scrapie infection (60 days, compared with 160 days in wild-type mice). However, the brains of these mice display remarkably little gliosis, no significant neuronal loss and very little accumulation of PrP^{Sc} [22,22a,23]. In contrast, hemizygous mice carrying only one functional copy of the *Prnp* gene, and expressing approx. 50% of the PrP^C found in wild-type mice, present with a relatively mild clinical course despite severe gliosis and more prominent neuronal loss.

A straightforward explanation for the above findings would be that a minimum period of time is required for PrP^{Sc} to elicit neuronal impairment. On the other hand, the observation that PrP^C expression levels (rather than the accumulation of PrP^{Sc}) correlate with the rate of clinical disease development suggests that neuronal damage may be the consequence of the conversion of PrP^C into PrP^{Sc} or some other agent.

In order to clarify the importance of these events, we again used grafting technology. Inoculated grafts developing a spongiform encephalopathy were kept viable for a period of time corresponding to seven times the lifespan of an infected donor mouse. This model enabled us to study the pathological chain of events leading to neurodegeneration, and to determine the end point of the disease. Of particular interest was the fate of neurons and astrocytes in scrapie-infected and degenerating grafts.

The first morphological changes in grafts characteristic of scrapie were detected 3 months after infection (Figure 3). At this time, the neuropil displayed small vacuoles, a diffusely reduced granular synaptophysin immunoreactivity, and significant gliosis. Later (5–9 months after infection), large vacuoles and a dense accumulation of large synaptophysin granules around the neuron cell bodies were seen. At late stages (more than 9 months after inoculation), the grafts were almost completely devoid of neurons, while the main cell population was formed by densely clustered astrocytes.

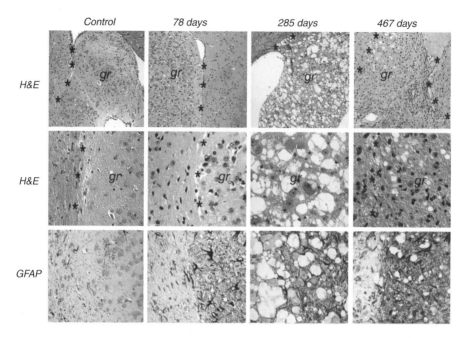

Figure 3. Progression of neurodegeneration in Scrapie-infected grafts
Grafts at different stages as defined by morphological criteria are ordered in columns, and different stains are ordered in rows. H&E, haematoxylin and eosin stain. Upper row, overview at low
magnification; lower rows, details at higher magnification. First column: control grafts typical of
an uninfected neural graft overexpressing PrP in a Prnp[o/o] host brain analysed 232 days after grafting. In the GFAP immunostain, only a few astrocytes appear in the graft (gr) and in the adjacent
host caudoputamen. The border between graft and recipient brain is indicated by asterisks.
Second column: 78 days after infection with scrapie (early stage). Mild but significant spongiosis
and brisk gliosis with the occurrence of paired astrocytes is observed. Note the restriction of
pathological changes to the graft. The diameter of vacuoles is approx. 0.5–2.0 μm. Third column:
285 days after scrapie infection (intermediate stage). Extreme spongiform changes have occurred,
with giant vacuoles leading to dramatic ballooning of neurons and the occurrence of astrocytes
with small, thickened processes. The diameter of the vacuoles is up to 40 μm. Fourth column:
467 days after infection (late stage). The cellular density has increased dramatically, with subtotal
loss of neuropil in the graft and preservation of small clustered astrocytes. Due to neuronal loss,
only a few small vacuoles are detected in such late-stage grafts. Magnification: upper row, ×40;
lower two rows, ×140.

Since BBB function relies on the interaction of astrocyte processes with
the CNS endothelium, we investigated the functionality of the BBB during
disease progression by means of gadolinium-enhanced MRI. The first MRI
scan was obtained 2–3 months after the transplantation procedure, and
showed full reconstitution of the BBB. Next, seven mice were inoculated
intracerebrally with mouse prions and four mice with mock inoculum. All
mice were scanned 2 months and 7 months later.

In seven mice with a tight BBB within the neural graft, scrapie infection
resulted in a persistent disruption of the BBB (Figures 4b, 4e and 4h), while no
BBB leakage was observed in mock-infected grafted control mice (Figures 4c,
4f and 4i). *Post mortem* analysis of the inoculated brains by immuno-

histochemistry confirmed the presence of a graft with accumulation of proteinase-resistant protein (Figures 4j and 4k), indicating that scrapie infection had taken place.

In contrast, further control mice with clinical symptoms of scrapie showed no detectable BBB impairment. This is in contrast with previous findings in hamsters, which exhibited patchy BBB leakage upon gadolinium-enhanced MRI [35], and in MRI studies of patients with CJD, which revealed increased signals in T2 weighted sequences (in which tissues with high water content appear bright) but no significant BBB disruption. These discrepancies may be ascribed to various factors. Since a functional BBB relies on intact astrocytic function and the interaction of astrocytic processes with blood vessel endothelia, the observed leakage in the transplantation model may be due to astrocytic damage. Indeed, strong gliosis occurs in the grafts 8–10 weeks after inoculation (Figure 3).

Perhaps scrapie affects the functional interaction of astrocytic end-feet with the vessel endothelium. It is known that this interaction contributes to BBB function. Interestingly, this putative functional impairment does not lead to the significant degeneration of astrocytes, which survive in large numbers until very late stages of disease in neural grafts. This is reflected by the persistence of BBB dysfunction in the grafting model during the incubation period, and is probably proportional to the extent of astrogliosis in the affected brain area. The very short latency between the onset of clinical symptoms and death, in combination with relatively mild gliosis, might explain why BBB dysfunction is not detectable in, for example, *tga20* mice. In those models of scrapie in which gliosis is more prominent (e.g. the hamster, and our grafting model), astrocyte dysfunction becomes prominent and is detectable as BBB leakage by gadolinium-enhanced MRI.

The findings described here confirm several predictions about the formal pathogenesis of spongiform encephalopathies: (i) scrapie leads to the selective loss of neurons; (ii) astrocytes and perhaps other neuroectodermal cells, while being affected by the disease, can survive and maintain their phenotypic characteristics for very long periods of time; and (iii) proteinase-resistant protein accumulates in neuroectodermal tissue at all stages of the disease. It is unclear whether proteinase-resistant PrP accumulates due to continued production by the remaining cells (mainly astrocytes) or whether previously accumulated PrP is not cleared after PrP production has ceased. Recent transgenic studies [33] with mice expressing hamster PrP^C under the control of a GFAP (the astrocyte intermediate-filament glial fibrillary acidic protein) promoter fragment have shown that astrocytes can support PrP^{Sc} accumulation, replication of the scrapie agent and induction of the disease.

Disruption of BBB function has been reported in experimental hamster scrapie [35], but has not been found in human spongiform encephalopathies. The localized BBB disruption in chronically infected grafts may contribute to the spread of prions from grafts to the surrounding brain which has been

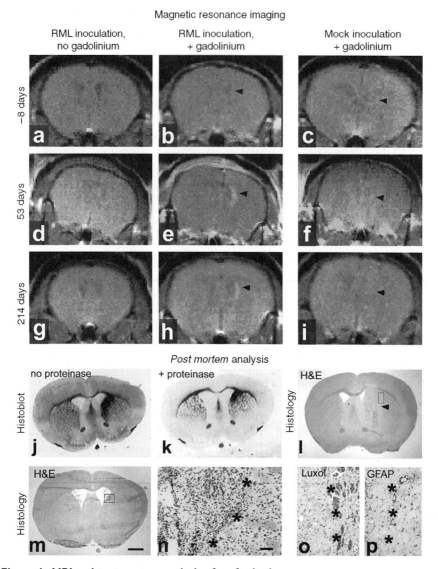

Figure 4. MRI and *post mortem* analysis of grafted mice
(a–c) MRI examination 78 days after grafting and 8 days before inoculation; (d–f) MRI at 53 days
after inoculation; (g–i) MRI at 214 days after inoculation. (a, d, g) MRI before and after Rocky
Mountain laboratory strain (RML) inoculation, before administration of gadolinium contrast medium; (b, e, h), same animal after application of gadolinium contrast medium; (c, f, i) contrast-
enhanced MRI of mock-infected control animal. No BBB leakage is detectable. *Post mortem* analysis
of the respective brains is shown in the lower panels. (j, k) Histoblot analysis of the RML-inoculat-
ed grafted brain shows strong PrP immunoreactivity in the area of the graft and the surrounding
brain. Following proteinase K digestion (k), PrP^{Sc} accumulation in the graft and in the surrounding
white matter is visualized. Histological analysis (m) of an adjacent frozen section [haema-
toxylin/eosin (H&E) stain] confirms the presence of a graft in the right ventricle (scale
bar = 1 mm). The area indicated by a box is shown at high-power magnification in (n): due to neuronal loss and astrogliosis, the graft becomes hypercellular (scale bar = 100 μm). (l) Paraffin histology of the grafted, non-inoculated brain reveals a large intraventricular graft. The area indicated by
a box in (l) is shown at high-power magnification stained with Luxol/Nissl (o) and immunostained
with GFAP (p). The graft is almost seamlessly integrated into the caudoputamen of the host, and
shows no spongiosis or gliosis. * Indicates the border between host and graft tissue.

described previously [22,22a]. This would account for the accumulation of proteinase-resistant PrP within the white matter and in brain areas surrounding the grafts. In view of the findings reported here, bulk-flow diffusion may be a mechanism contributing to prion spread as a consequence of prion-induced disruption of the BBB.

Summary

- *For the study of prion neurotoxicity, we used neural-grafting techniques: mice devoid of the normal host prion protein (Prnp^{o/o} mice) received a neural graft and were intracerebrally infected with mouse prions.*
- *The growth and differentiation properties of neural grafts were defined. Growth of embryonic neuroectodermal tissue was optimal at gestational days 12.5–13.5. The blood–brain barrier is reconstituted after 7 weeks in most animals.*
- *Scrapie-infected PrP^C-expressing grafts develop a severe spongiform encephalopathy and contain proteinase-resistant protein and infectivity.*
- *Infected grafts deliver high amounts of prions to the host brain without eliciting disease.*
- *Infected grafts show a progressive disruption of the blood–brain barrier.*
- *Following intraocular prion inoculation of a transplanted Prnp^{o/o} mouse, prions do not reach the intracerebral graft, indicating that PrP expression is required for propagation along the optic tract.*

We thank Andrea Burlet and Marianne König for histotechnical help, and Norbert Wey for photographic work. The work described is being supported by the Kanton of Zürich, by grants from the Schweizerischer Nationalfonds and Nationales Forschungsprogramm NFP38 (to A.A. and Charles Weissmann), the European Union, the Bundesämter für Gesundheit and für Veterinärwesen and the Migros Foundation (to A.A.), and by postdoctoral fellowships from EMBO to C.M. and from Deutsche Forschungsgemeinschaft to M.A.K.

References

1. Aguzzi, A. & Weissmann, C. (1997) Prion research: the next frontiers. *Nature (London)* **389**, 795–798
2. Weissmann, C. & Aguzzi, A. (1997) Bovine spongiform encephalopathy and early onset variant Creutzfeldt–Jakob disease. *Curr. Opin. Neurobiol.* **7**, 695–700
3. Brüstle, O., Spiro, A.C., Karram, K., Choudhary, K., Okabe, S. & McKay, R.D. (1997) In vitro-generated neural precursors participate in mammalian brain development. *Proc. Natl. Acad. Sci. U.S.A.* **94**, 14809–14814
4. Dunnett, S.B. (1990) Neural transplantation in animal models of dementia. *Eur. J. Neurosci.* **2**, 567–587

5. Büeler, H.R., Aguzzi, A., Sailer, A., Greiner, R.A., Autenried, P., Aguet, M. & Weissmann, C. (1993) Mice devoid of PrP are resistant to scrapie. *Cell* **73**, 1339–1347

6. Games, D., Adams, D., Alessandrini, R., Barbour, R., Berthelette, P., Blackwell, C., Carr, T., Clemens, J., Donaldson, T., Gillespie, F. et al. (1995) Alzheimer-type neuropathology in transgenic mice overexpressing V717F beta-amyloid precursor protein. *Nature (London)* **373**, 523–527

7. LaFerla, F.M., Tinkle, B.T., Bieberich, C.J., Haudenschild, C.C. & Jay, G. (1995) The Alzheimer's A beta peptide induces neurodegeneration and apoptotic cell death in transgenic mice. *Nature Genet.* **9**, 21–30

8. Klein, R., Smeyne, R.J., Wurst, W., Long, L.K., Auerbach, B.A., Joyner, A.L. & Barbacid, M. (1993) Targeted disruption of the trkB neurotrophin receptor gene results in nervous system lesions and neonatal death. *Cell* **75**, 113–122

9. Magyar, J.P., Bartsch, U., Wang, Z.Q., Howells, N., Aguzzi, A., Wagner, E.F. & Schachner, M. (1994) Degeneration of neural cells in the central nervous system of mice deficient in the gene for the adhesion molecule on Glia, the beta 2 subunit of murine Na,K-ATPase. *J. Cell Biol.* **127**, 835–845

10. Smeyne, R.J., Klein, R., Schnapp, A., Long, L.K., Bryant, S., Lewin, A., Lira, S.A. & Barbacid, M. (1994) Severe sensory and sympathetic neuropathies in mice carrying a disrupted Trk/NGF receptor gene. *Nature (London)* **368**, 246–249

11. Bladt, F., Riethmacher, D., Isenmann, S., Aguzzi, A. & Birchmeier, C. (1995) Essential role for the c-met receptor in the migration of myogenic precursor cells into the limb bud. *Nature (London)* **376**, 768–771

12. Meyer, D. & Birchmeier, C. (1995) Multiple essential functions of neuregulin in development. *Nature (London)* **378**, 386–390

13. Gassmann, M., Casagranda, F., Orioli, D., Simon, H., Lai, C., Klein, R. & Lemke, G. (1995) Aberrant neural and cardiac development in mice lacking the ErbB4 neuregulin receptor. *Nature (London)* **378**, 390–394

14. Isenmann, S., Brandner, S. & Aguzzi, A. (1996) Neuroectodermal grafting: a new tool for the study of neurodegenerative diseases. *Histol. Histopathol.* **11**, 1063–1073

14a. Isenmann, S., Brandner, S., Sure, U. & Aguzzi, A. (1996) Telencephalic transplants in mice: characterization of growth and differentiation patterns. *Neuropathol. Appl. Neurobiol.* **21**, 108–117

15. Isenmann, S., Molthagen, M., Brandner, S., Bartsch, U., Kühne, G., Magyar, J.P., Sure, U., Schachner, M. & Aguzzi, A. (1995) The AMOG AMOG/b2 subunit of Na,K-ATPase is not necessary for long term survival of telencephalic grafts. *Glia* **15**, 377–388

16. Rosenstein, J.M. & Brightman, M.W. (1983) Circumventing the blood–brain barrier with autonomic ganglion transplants. *Science* **221**, 879–881

17. Stewart, P.A. & Wiley, M.J. (1981) Developing nervous tissue induces formation of blood–brain barrier characteristics in invading endothelial cells: a study using quail–chick transplantation chimeras. *Dev. Biol.* **84**, 183–192

18. Bertram, K.J., Shipley, M.T., Ennis, M., Sanberg, P.R. & Norman, A.B. (1994) Permeability of the blood–brain barrier within rat intrastriatal transplants assessed by simultaneous systemic injection of horseradish peroxidase and Evans blue dye. *Exp. Neurol.* **127**, 245–252

19. Sailer, A., Büeler, H., Fischer, M., Aguzzi, A. & Weissmann, C. (1994) No propagation of prions in mice devoid of PrP. *Cell* **77**, 967–968

20. Büeler, H.R., Fischer, M., Lang, Y., Bluethmann, H., Lipp, H.P., DeArmond, S.J., Prusiner, S.B., Aguet, M. & Weissmann, C. (1992) Normal development and behaviour of mice lacking the neuronal cell-surface PrP protein. *Nature (London)* **356**, 577–582

21. Forloni, G., Angeretti, N., Chiesa, R., Monzani, E., Salmona, M., Bugiani, O. & Tagliavini, F. (1993) Neurotoxicity of a prion protein fragment. *Nature (London)* **362**, 543–546

22. Brandner, S., Isenmann, S., Raeber, A., Fischer, M., Sailer, A., Kobayashi, Y., Marino, S., Weissmann, C. & Aguzzi, A. (1996) Normal host prion protein necessary for scrapie-induced neurotoxicity. *Nature (London)* **379**, 339–343

22a. Brandner, S., Raeber, A., Sailer, A., Blaettler, T., Fischer, M., Weissmann, C. & Aguzzi, A. (1996) Normal host prion protein (PrP^C) required for scrapie spread within the central nervous system. *Proc. Natl. Acad. Sci. U.S.A.* **93**, 13148–13151

23. Fischer, M., Rülicke, T., Raeber, A., Sailer, A., Moser, M., Oesch, B., Brandner, S., Aguzzi, A. & Weissmann, C. (1996) Prion protein (PrP) with amino-proximal deletions restoring susceptibility of PrP knockout mice to scrapie. *EMBO J.* **15**, 1255–1264

24. Shyng, S.L., Huber, M.T. & Harris, D.A. (1993) A prion protein cycles between the cell surface and an endocytic compartment in cultured neuroblastoma cells. *J. Biol. Chem.* **268**, 15922–15928

25. Fraser, H. (1982) Neuronal spread of scrapie agent and targeting of lesions within the retinotectal pathway. *Nature (London)* **295**, 149–150

26. Prusiner, S.B., Groth, D., Serban, A., Koehler, R., Foster, D., Torchia, M., Burton, D., Yang, S.L. & DeArmond, S.J. (1993) Ablation of the prion protein (PrP) gene in mice prevents scrapie and facilitates production of anti-PrP antibodies. *Proc. Natl. Acad. Sci. U.S.A.* **90**, 10608–10612

27. Borchelt, D.R., Koliatsos, V.E., Guarnieri, M., Pardo, C.A., Sisodia, S.S. & Price, D.L. (1994) Rapid anterograde axonal transport of the cellular prion glycoprotein in the peripheral and central nervous systems. *J. Biol. Chem.* **269**, 14711–14714

28. Kimberlin, R.H. & Walker, C.A. (1986) Pathogenesis of scrapie (strain 263K) in hamsters infected intracerebrally, intraperitoneally or intraocularly. *J. Gen. Virol.* **67**, 255–263

29. Scott, J.R. & Fraser, H. (1989) Enucleation after intraocular scrapie injection delays the spread of infection. *Brain Res.* **504**, 301–305

30. Collinge, J., Whittington, M.A., Sidle, K.C., Smith, C.J., Palmer, M.S., Clarke, A.R. & Jefferys, J.G. (1994) Prion protein is necessary for normal synaptic function. *Nature (London)* **370**, 295–297

31. Aguzzi, A. (1997) Neuro-immune connection in spread of prions in the body? *Lancet* **349**, 742–743

32. Masters, C.L. & Richardson, E.P. (1978) Subacute spongiform encephalopathy (Creutzfeldt–Jakob disease): the nature and progression of spongiform change. *Brain* **101**, 333–344

33. Raeber, A.J., Race, R.E., Brandner, S., Priola, S.A., Sailer, A., Bessen, R.A., Mucke, L., Manson, J., Aguzzi, A., Oldstone, M.B.A. et al. (1997) Astrocyte-specific expression of hamster prion protein (PrP) renders PrP knockout mice susceptible to hamster scrapie. *EMBO J.* **16**, 6057–6065

34. Jeffrey, M., Fraser, J.R., Halliday, W.G., Fowler, N., Goodsir, C.M. & Brown, D.A. (1995) Early unsuspected neuron and axon terminal loss in scrapie-infected mice revealed by morphometry and immunocytochemistry. *Neuropathol. Appl. Neurobiol.* **21**, 41–49

35. Chung, Y.L., Williams, A., Beech, J.S., Williams, S.C., Bell, J.D., Cox, I.J. & Hope, J. (1995) MRI assessment of the blood–brain barrier in a hamster model of scrapie. *Neurodegeneration* **4**, 203–207

12

Pathological mechanisms in Huntington's disease and other polyglutamine expansion diseases

Astrid Lunkes[*], Yvon Trottier[*] and Jean-Louis Mandel[*][†][1]

[*]Institut de Génétique et Biologie Moléculaire et Cellulaire, INSERM/CNRS/Université Louis Pasteur, BP 163, 67 404 Illkirch Cedex, CU de Strasbourg, and [†]Hopitaux Universitaires de Strasbourg, Strasbourg, France

Introduction

Since 1991, eleven monogenic diseases have been shown to be caused by unstable expansions of trinucleotide repeats. More recently, the scope of repeat-expansion diseases was enlarged by the discovery of an unstable expansion of a dodecamer repeat, and of diseases caused by stable expansions [1,2] (Figure 1). Detailed reviews of individual diseases and the mutation mechanisms involved can be found in a recent book [3].

The unstable-expansion diseases can be divided into two broad classes, depending on whether the repeat is located in non-coding or coding regions of the affected gene (Table 1). In the first class, expansions are generally massive (from about 300 bp to several thousand base pairs) and affect the expression of the gene in both neural and non-neural tissues. Thus, while these diseases can be broadly defined as neurological diseases, clinical symptoms in most cases affect a number of organs. The second class is more homogeneous, and includes seven purely neurodegenerative diseases caused by unstable but mod-

[1]To whom correspondence should be addressed.

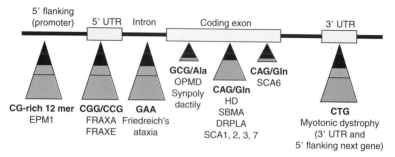

Figure 1. Type and location of repeat expansions linked to diseases
The size of the triangles reflects the size of the expansions found in the diseases. In the triangles, black indicates normal alleles, grey indicates unstable pre-mutations and blue indicates pathological alleles. UTR, untranslated region. For definitions of disease abbreviations, see Table 1.

erate expansion of a CAG repeat coding for a polyglutamine stretch in the expressed protein. These disorders include Huntington's disease (HD) and various spinocerebellar ataxias (SCAs). Their pathogenic mechanism has been the object of intensive recent studies, and these will be summarized in this essay.

Above a threshold of about 35–50 trinucleotide repeats (or 12 dodecamer repeats, corresponding to a similar length of 100–150 bp), repeats that are linked to diseases become unstable and tend to expand further in successive generations. The length of the repeat is correlated with the age of onset of the disease (or, in the case of fragile X syndrome, with the risk of having affected children), and the tendency to expand depends strongly on the sex of the transmitting parent. In autosomal dominant diseases, this will result in a tendency towards an earlier age of onset and increased severity in successive generations; this phenomenon is called anticipation, and is more marked in myotonic dystrophy and SCA7 than in other diseases. In the case of X-linked fragile X syndrome, anticipation is manifested by an increased risk of developing the disease in successive generations upon maternal transmission. Anticipation cannot be observed in autosomal recessive diseases such as Friedreich's ataxia or progressive myoclonic epilepsy.

The small group of diseases that are due to stable coding expansions includes SCA6 and oculopharyngeal muscular dystrophy (Table 1). The stable expansion of a polyalanine stretch affecting a homeobox gene (*HOXD13*) that is responsible for a limb malformation (synpolydactily) has a different mutation mechanism as it affects an impure repeat, and is probably caused by unequal crossover.

HD: clinical features

HD is the most common and the best known of the polyglutamine expansion diseases. It has a prevalence of about 1 in 12000–15000 in Western European populations [4]. The disease begins gradually, leading to mood disturbances,

Table 1. Diseases associated with repeat expansions

(A) Unstable expansions

Class I: massive non-coding expansions

Repeat motif	Disease	Transmission	Protein involved and localization of repeat	Effect	Conventional mutations	Parental bias of expansion
CGG (methylated)	Fragile X mental retard-ation syndrome (FRAXA)	XD	FMR1 (RNA binding protein); 5' UTR	LOF (transcription)	Same disease	Maternal
CCG (methylated)	FRAXE (mild mental retardation)	XR	FMR2 (transcription factor?); 5' UTR or 5' flanking sequence	LOF (transcription)	Same disease	Maternal
CTG	Myotonic dystrophy	AD	DMPK (serine/threonine kinase); 3' UTR and DMAHP (homeoprotein); 5' flanking sequence	LOF? (haploid insufficiency) and/or toxic RNA?	No	50–100 paternal >maternal; 100–500 maternal ≈paternal; >>500 maternal
GAA	Friedreich's ataxia	AR	Frataxin (mitochondrial protein); intronic	LOF (transcription?)	Same disease	Maternal?
CG-rich dodecamer (?)	Progressive myoclonus epilepsy (EPM1)	AR	Cystatin B (proteinase inhibitor); 5' flanking sequence	LOF (transcription)	Same disease	Paternal?

(contd.)

Table 1. (Contd.)

Class II: moderate CAG/polyglutamine expansions

Repeat motif	Disease	Transmission	Protein involved and localization of repeat	Effect	Conventional mutations	Parental bias of expansion
	Spinal and bulbar muscular atrophy (SBMA or Kennedy disease)	XR	Androgen receptor	GOTP	Testicular feminization syndrome/androgen insensitivity	Paternal (weak)
	Huntington's disease (HD)	AD	Huntingtin	GOTP	No	Paternal
	Dentatorubral–pallido-luysian atrophy (DRPLA)	AD	Atrophin	GOTP	No	Paternal
	Spinocerebellar ataxia 1 (SCA1)	AD	Ataxin 1	GOTP	No	Paternal
	SCA2	AD	Ataxin 2	GOTP	No	Paternal >maternal
	SCA3/Machado–Joseph disease (MJD)	AD	Ataxin 3	GOTP	No	Paternal >maternal
	SCA7 (with macular degeneration)	AD	Ataxin 7	GOTP	No	Paternal >maternal

(contd.)

Table 1. (Contd.)

(B) Stable coding expansions

Repeat motif	Disease	Transmission	Protein involved and localization of repeat	Effect	Conventional mutations	Parental bias of expansion
CAG/Gln	SCA6	AD	α1A calcium channel	Misfunction?	Familial hemiplegic migraine/episodic ataxia type 2	(Stable)
GCG/Ala	Oculopharyngeal muscular atrophy (OPMD)	AD	Poly(A) binding protein 2 (PABP2)	GOTP?	No	(Stable)
GCN/Ala (impure)	Synpolydactily	AD	HOXD13 (homeobox gene)	Dominant negative or GOTP?	No	(Stable)

The large expansions associated with fragile sites FRAXF, FRA11B, FRA16A (methylated CGG/CCG) and FRA16B (33-mer; AT-rich) have been omitted, as they do not appear to cause a disease directly. Transmission is X-linked (X) or recessive (R). The column headed 'Conventional mutations' indicates whether or not conventional mutations in the same gene have been found to cause the same disease as the expansion or a different disease (as indicated); 'no' means that no conventional pathological mutations have been found in this gene. Abbreviations: UTR, untranslated region; LOF, loss of function; GOTP, gain of toxic property at the protein level, correlated with the presence of nuclear inclusions.

involuntary movements (chorea) and cognitive impairment, and progresses inexorably to death 15–20 years after onset [5]. The motor disability affects both involuntary and voluntary movements. Chorea is observed in approximately 90% of patients, and dystonia becomes a prominent feature in later stages of the illness. A progressive loss of co-ordination of voluntary movements leads to a state in which voluntary movement is impossible. Cognitive disorder begins with a loss of mental flexibility and a slowing of intellectual processes, and progresses to profound dementia. Mood abnormalities often appear a few years before the onset of the movement disorder. Age at onset of symptoms is classically between 30 and 50 years; however, juvenile and late-onset cases also occur. The juvenile forms (onset before the age of 20) represent about 6–10% of cases, and are almost exclusively transmitted by the father. The disease is inherited dominantly, with almost complete penetrance (see below), and homozygous and sporadic cases are rare. The fact that the rare homozygous patients do not show greater severity provided a decisive argument in favour of a gain of function conferred by the mutation, since dominant diseases caused by a loss-of-function mutation are much more severe in the homozygous state [6].

HD: neuropathology

The clinical progression of HD is paralleled by brain degeneration. While the caudate nucleus and putamen show the most dramatic changes, there is an overall atrophy of the brain. In advanced cases, the brain often weighs 20–25% less than normal. Within the striatum, there is selective loss of medium spiny GABAergic neurons (GABA is γ-aminobutyric acid), but a relative preservation of interneurons. The overall architecture of the caudate nucleus and the adjacent putamen is destroyed, and gliosis may be prominent. Prior to cell death, neuronal dysfunction is manifested by anomalies of dendritic endings and changes in the density and shape of the spines. Extensive neuronal loss also occurs in the deep layers of the cerebral cortex, and may affect other brain structures, such as the thalamus and the hypothalamic lateral nucleus.

Expansion mutation and genotype–phenotype correlations

The genetic defect causing HD is an intragenic expansion of a CAG repeat that leads to the elongation of a polyglutamine stretch in the N-terminus of the target protein [7]. The *HD* gene encodes a very large protein (350 kDa) of unknown function, named huntingtin. The first exon of the gene contains a CAG repeat that is polymorphic in the normal population, with between 6 and 35 units being present. In HD patients, the CAG repeat is expanded to between 12 and 36 units. There is an inverse correlation between the length of the expansion and age of onset (the repeat size of the mutated allele accounts for about 70% of the variation in age of onset) [6,8,9]. Most mutated alleles contain between 40 and 55 repeats, and these are associated with onset in middle age [6]. When between 36 and 39 repeats are present, the penetrance of the disease is not complete, and some carriers of such alleles may remain healthy until about 80 years of age [8,9a]. Patients with 60 or more repeats present a juvenile onset. Although the correlation between the length of the expansion and age of onset is well established, the wide confidence interval renders this correlation of little use for predicting the age of onset in individuals. The identification of other putative genetic or environmental factors that may account for some of the variation in age of onset would be of considerable interest, as it could provide additional targets for therapeutic intervention. A search for candidate modifier genes has been initiated [9].

While the mutated alleles show slight instability in somatic tissues, marked instability is observed in male germ cells (sperm). Here the size of the CAG repeat is often very heterogeneous, and greater than in white blood cells from the same subject. This accounts for the paternal bias of transmission of the juvenile forms. In such cases, the father transmits a mutated *HD* allele to his child with an increase in length of about 10 or more CAG repeats, leading to anticipation in the age of onset (i.e. the child will become affected at a much earlier age than the father, and in rare cases may even develop the disease while the father is still clinically unaffected). However, in most cases of paternal transmission, allelic instability is rather limited, and thus will have a limited impact on the child's age of onset. In cases of maternal transmission, the mutated alleles remain quite stable, with an average change of +0.6 CAG repeats, compared with +6 for paternal transmission. Although sporadic cases are rare, they may occur due to new germline mutations arising at low frequency in males carrying an allele in the high normal range (30–35 CAGs) by an increase in the repeat size into the pathological range. This mutation may then be transmitted to subsequent generations.

Huntingtin is a cytoplasmic protein found in many peripheral tissues, but it is particularly abundant in neurons. Its wide expression in the brain contrasts with the selectivity of neuronal loss in HD [10]. In the brain, huntingtin is localized in part to nerve endings, suggesting a role in synaptic vesicle transport. While its exact function remains to be elucidated, huntingtin has an important role during early embryonic development. Indeed, mouse embryos homozygous for inactivating mutations in the *HD* gene cannot complete gastrulation. Mice with greatly decreased *HD* gene expression have been generated, and these show abnormal brain development [11].

Polyglutamine expansions in other neurodegenerative disorders

The number of neurodegenerative diseases identified as being caused by an expansion of a CAG/polyglutamine repeat has grown steadily since 1991. The last to be added to the list is a type of spinocerebellar ataxia, SCA7, which is the only one of these diseases that also affects the retina [12]. As with HD, all

these diseases show a strong inverse correlation between age of onset of clinical symptoms and the length of the abnormal repeat [6]. The pathological
threshold in four of these diseases (35–40 glutamines) is strikingly similar to
that for HD (Table 2) [12a]. Below this threshold, the high degree of polymorphism of the repeat observed in the normal population indicates that these
proteins tolerate a wide variation in polyglutamine length without adverse

Table 2. CAG trinucleotide repeat diseases

Disease	Sites of neuropathology	Repeat number Normal	Repeat number Disease	Intracellular localization of encoded protein
SBMA	Motor neurons (anterior horn cells, bulbar neurons) and dorsal root ganglia	11–34	40–62	Nuclear (androgen receptor)
DRPLA	Globus pallidus, dentato–rubral and subthalamic nucleus	7–35	49–88	Cytoplasmic
HD	Striatum (medium spiny neurons) and cortex in late stage	6–35	36–121	Cytoplasmic
SCA1	Cerebellar cortex (Purkinje cells), dentate nucleus and brainstem	6–39	40–81	Nuclear, cytoplasmic
SCA2	Cerebellum, pontine nuclei, substantia nigra	15–29	35–64	Cytoplasmic
SCA3	Substantia nigra, globus pallidus, pontine nucleus, cerebellar cortex	13–42	61–84	Cytoplasmic
SCA6	Cerebellar and mild brainstem atrophy	4–18	21–30	Transmembrane (calcium channel subunit α1A)
SCA7	Photoreceptors and bipolar cells, cerebellar cortex, brainstem	7–17	37–130	Nuclear

For definitions of abbreviations, refer to Table 1. SCA6 shows the lowest pathological range, and
comparison with other mutations in the same gene suggests that polyglutamines affect normal
function.

effect. All eight diseases are characterized by slowly progressing neuronal degeneration in specific regions of the brain, which differ (but with some overlap) between the various diseases (Table 2). With the exception of SCA6 [13], this specificity of neuronal death shows no obvious correlation with the rather ubiquitous pattern of expression of the affected protein.

The functions of these proteins are unknown in six of the eight disorders. The two exceptions are the androgen receptor implicated in spinal and bulbar muscular atrophy (SBMA; also known as Kennedy disease), and a calcium channel subunit mutated in SCA6. Except for their polyglutamine tracts, the proteins do not resemble each other. The genetic features of SBMA (the first polyglutamine expansion disease identified in 1991) indicate that, as in HD, the expansion causes a gain of toxic property, and not a loss of function. Mutations that cause a complete loss of function of the androgen receptor gene are responsible for the testicular feminization syndrome, which has no neurological symptoms. By analogy, the same gain of toxic property is also expected to occur in the other diseases, with the likely exception of SCA6. The latter disease is characterized by a much lower pathological threshold in polyglutamine size and, most importantly, conventional loss-of-function mutations in the affected gene result in acetazolamide-responsive episodic ataxia, a disease that shares overlapping clinical features with SCA6 and leads to progressive cerebellar degeneration in some subjects [13]. The finding that the abnormal proteins present in patients with HD, SCA1, SCA2, SCA3 and SCA7 are selectively recognized by a monoclonal antibody provides additional support for a common pathogenic mechanism, suggesting an abnormal conformation of the elongated polyglutamines [14]. Some tantalizing questions are raised by these observations. For example, how do these elongated polyglutamines cause neuronal death, and what features can account for the selectivity with which various neuronal populations are affected?

Nuclear inclusions and mechanisms of neurodegeneration

A mouse model for HD has been described in which expression of a very short truncated huntingtin fragment carrying between 115 and 155 glutamine residues (corresponding to a disease onset in the first years of life) leads to a strong neurological phenotype and significant brain shrinkage [15]. In some, but not all, areas of the brains of these mice, Davies et al. [16] reported the conspicuous presence of nuclear inclusions that stained with an anti-huntingtin antibody and also showed ubiquitin immunoreactivity. In some regions (corresponding to those affected in HD), it was estimated that most neurons contained a nuclear inclusion. A similar observation was made in another mouse model, in which overexpression in Purkinje cells of a full-length ataxin 1 (see Table 1) containing 82 glutamines (also resulting in infantile onset) leads to severe ataxia, cerebellar dysfunction and ultimately Purkinje cell death [17]. In both mouse models, the appearance of the nuclear inclusions preceded the onset of clinical symptoms by several weeks.

Since each mouse model has an artificial feature (i.e. expression of a truncated protein [15] or gross overexpression of the full-length protein [17]), it was most important to check whether similar observations can be made in patients. This has now been done for four diseases: HD, SCA1, SCA3 and dentatorubral–pallidoluysian atrophy (DRPLA) [18–21]. In HD, inclusions are observed in the cortex and in medium-sized neurons of the striatum, but not in unaffected brain areas or in controls. These nuclear inclusions are stained by an antibody specific for the N-terminal region of huntingtin, lying very close to the polyglutamine tract, and by an anti-ubiquitin antibody, but not by antibodies recognizing other regions of huntingtin [18,19]. In one study, the nuclear inclusions appeared much rarer in adult-onset cases; here another type of pathological immunoreactive aggregate was observed that looked like dystrophic neurites [19]. However, the latter findings were not replicated in another study [18]. Nuclear inclusions have also been observed in affected brain areas of SCA3/Machado–Joseph disease (MJD) patients [20] and of a juvenile SCA1 patient [17]. At first sight it is surprising that huntingtin or ataxin 3, which are normally cytoplasmic proteins, appear to aggregate in the nucleus. At least in the case of huntingtin, only a short N-terminal fragment accumulates in the inclusions [18,19]. Thus processing of the mutated protein may lead to the production of such a fragment, which would diffuse passively or be transported actively to the nucleus, where the environment would favour aggregation. Subsequent ubiquitination would reflect a failed attempt to degrade the inclusions.

Current observations suggest that nuclear inclusions cause neuronal dysfunction, and neuronal death may be only a late event. This is difficult to study in humans, as most samples available are from patients at the end of many years of disease progression. The mouse models are thus invaluable. In SCA1 mice, the ataxia phenotype and decreased arborization of Purkinje cells occur long before the loss of Purkinje cells [22]. In the HD mouse model, the neurological symptoms are very severe, but there are no obvious signs of neuronal death or widespread astrogliosis. The 20% brain weight loss observed in these mice is currently unexplained [16].

Further studies of the protein mutated in SCA1 give possible clues about pathological mechanisms and their neuronal specificity. Using the yeast two-hybrid system to screen for proteins interacting with ataxin 1, Matilla et al. [23] have found a leucine-rich protein (leucine-rich acidic nuclear protein; LANP) that interacts much better with a mutated ataxin 1 with 82 repeats than with normal ataxin with only 30 glutamines. LANP is expressed predominantly in Purkinje cells (the primary pathological target in SCA1) and co-localizes with ataxin 1 in the nuclear matrix. Thus mutated ataxin may interfere with the normal function of LANP [23]. Another indication that the nuclear matrix may be affected is provided by the abnormal distribution of the PML protein (a nuclear matrix protein associated with nuclear bodies; the *PML* gene is fused to a retinoic acid receptor gene in acute promyelocytic leukaemia) in COS cells overexpressing mutated ataxin [17]. It remains to be seen whether other proteins showing both sensitivity to polyglutamine repeat length and appropriate neuronal specificity of expression can be found to interact with mutated proteins in other diseases, and may thus account, at least in part, for the selectivity of neuronal degeneration. Two proteins whose interactions with huntingtin are modulated by the number of glutamines have been described (huntingtin-associated protein 1 and huntingtin-interacting protein 1), but they do not show specificity of expression in the neurons selectively affected in HD [24,25]. However, nuclear inclusions and a neurological phenotype (but without obvious neurodegeneration) were obtained by inserting a long CAG/polyglutamine repeat within the mouse hypoxanthine phosphoribosyltransferase gene [26]. This gene is not implicated in a polyglutamine expansion disease in humans, suggesting that specific interacting proteins may not be mandatory in order to elicit polyglutamine-mediated neuronal dysfunction.

Other (not necessarily exclusive) explanations for the observed regional selectivity could lie in subtle differences in expression of the target proteins in different neuronal populations, or in alterations in the processing of these proteins to produce truncated fragments with a higher aggregation potential in some neurons. In support of the first hypothesis is the similarity in the regions affected in HD patients and in the mouse HD model that expresses a highly truncated protein under the control of the promoter region of the human HD gene [16]. In rats, the vulnerability of different neurons in the striatum appears to be correlated with a normal level of huntingtin expression [27]. It is well known that the kinetics of protein aggregation can be markedly influenced by rather small differences in protein concentration [28]. On the other hand, it has been suggested that the expanded polyglutamine repeat may lead to an increased sensitivity of huntingtin to cleavage by apoptosis-related proteases, such as caspase 3 [29].

How do the nuclear inclusions form, and what is the basis of the polyglutamine length sensitivity? Scherzinger et al. [30] found that a short truncated huntingtin with 51 glutamines may form amyloid-like aggregates *in vitro*, with a fibrillar morphology similar to those observed in Alzheimer's disease and prion diseases [28]. Indeed, Max Perutz and co-workers predicted that polyglutamine tracts may oligomerize and form β-pleated sheets [31], a type of structure also found in Alzheimer β-amyloid plaques and in prion deposits [28]. However, a search for amyloid deposits in the brains of HD patients was negative, implying that the nuclear inclusions are not found under the standard conditions for amyloid detection.

The threshold effect observed for the various polyglutamine diseases may be explained either by a gradual increase in aggregation propensity with polyglutamine length, or by a conformational change that occurs at the 35–40 glutamine threshold. In favour of the latter, a monoclonal antibody has been found that, in Western blots, recognizes specifically the mutated proteins, but not the normal ones (at least in HD and in SCA1, 2, 3 and 7) [14]. It will be important to obtain crystallographic data on the structure of long polyglutamine repeats and their aggregates, as this might be used to identify drugs interfering with this process. The aggregation properties of the polyglutamine tracts are also influenced by the nature of the affected protein. In an *in vitro* model, aggregation of the truncated huntingtin with 51 glutamines was inhibited by fusion to an unrelated protein. Similarly, the mouse model overexpressing in Purkinje cells a truncated mutated ataxin 3 shows extensive cerebellar degeneration, whereas expression at the same level of the complete mutated protein does not cause a neurological phenotype [32]. This protective effect might explain the higher pathological threshold of 61 glutaminess in SCA3/MJD, or the innocuity of the tract of 38 glutamines carried in the transcription factor TBP by the majority of the human population [14]. This would also predict that proteolytic processing of the mutated proteins is an important factor in pathogenesis, in agreement with the finding of only a short N-terminal fragment of huntingtin in the nuclear inclusions of HD patients [18,19]. Indeed, we have recently constructed a cellular model in which inducible expression of full-length mutated huntingtin leads to cytoplasmic and nuclear inclusions that contain only an N-terminal segment of huntingtin [33].

It should be noted that the alternative (but not exclusive) hypothesis of the involvement of transglutaminases in the pathogenesis of polyglutamine diseases has recently gained some support from *in vitro* experiments. These enzymes catalyse the covalent cross-linking of glutamine residues to polypeptides containing lysyl groups, and elongated polyglutamines appear to be preferential substrates in the reaction *in vitro* [34]. Furthermore, inhibitors of transglutaminases, such as cystamine and monodansyl cadaverine, inhibit aggregate formation and cell death in cultured cells expressing a mutated and truncated DRPLA protein [21]. It remains to be shown whether transglutaminases indeed play a role in the formation of pathogenic aggregates in patients or mouse models.

In conclusion, HD and related diseases have joined the group of neurodegenerative diseases caused by amyloid-like aggregates, and this may lead to a cross-fertilization of research efforts. The creation and analysis of animal and cellular models for HD should lead to a better understanding of the pathogenetic mechanisms involved, and hopefully to the identification of therapeutic targets.

Summary

- *HD is an autosomal dominant neurodegenerative disorder characterized by involuntary movements, cognitive impairment progressing to dementia, and mood disturbances. The brains of patients show extensive neuronal loss in the striatum, and the cerebral cortex is also affected.*

- *The genetic defect causing HD is an expansion of a CAG repeat encoding a polyglutamine stretch in the target protein, named huntingtin.*

- *The age of onset of HD is inversely correlated with the size of the expansion.*

- *Polyglutamine expansion represents a novel cause of neurodegeneration, which has been shown to be responsible for seven other inherited disorders.*

- *The polyglutamine expansion confers a gain of toxic property to the mutated target proteins.*

- *Molecular and cellular studies of the brains of patients and of mice models of polyglutamine expansion diseases have led to the identification of abnormal intracellular inclusions representing aggregation of the mutated protein. However, the mechanism whereby such polyglutamine expansion leads to selective neuronal dysfunction and death is still puzzling.*

References

1.   Lalioti, M.D., Scott, H.S., Buresi, C., Rossier, C., Bottani, A., Morris, M.A., Malafosse, A. & Antonarakis, S.E. (1997) Dodecamer repeat expansion in cystatin B gene in progressive myoclonus epilepsy. *Nature (London)* **386**, 847–851

2.   Brais, B., Bouchard, J.-P., Xie, Y.-G., Rochefort, D.L., Chrétien, N., Tomé, F.M., Lafrenière, R.G., Rommens, J.M., Uyama, E., Nohira, O. et al. (1998) Short GCG expansions in the *PABP2* gene cause oculopharyngeal muscular dystrophy. *Nat. Genet.* **18**, 164–167

3.   Warren, S.T. & Wells, R.D. (1998) *Genetic Instabilities and Hereditary Neurological Diseases*, Academic Press, San Diego

4.   Harper, P.S. (1992) The epidemology of Huntington's disease. *Hum. Genet.* **89**, 365–376

5.   Harper, P.S. (1991) *Huntington's Disease*, W.B. Saunders, London

6.   Gusella, J.F., McNeil, S., Persichetti, F., Srinidhi, J., Novelletto, J., Bird, E., Faber, P., Vonsattel, J.P., Myers, R.H. & MacDonald, M.E. (1996) Cold Spring Harbor Symp. Quant. Biol. **61**, 615–626

7.   The Huntington's Disease Collaborative Research Group (1993) A novel gene containing a trinucleotide repeat that is expanded and unstable on Huntington's disease chromosomes. *Cell* **72**, 971–983

8.   Brinkman, R.R., Mezei, M.M., Theilmann, J., Almqvist, E. & Hayden, M.R. (1997) The likelihood of being affected with Huntington's disease by a particular age, for a specific CAG size. *Am. J. Hum. Genet.* **60**, 1202–1210

9.   Rubinsztein, D.C., Leggo, J., Chiano, M., Dodge, A., Norbury, G., Rosser, E. & Craufurd, D. (1997) Genotypes at the GluR6 kainate receptor locus are associated with variation in the age of onset of Huntington's disease. *Proc. Natl. Acad. Sci. U.S.A.* **94**, 3872–3876

9a. Rubinsztein, D.C., Leggo, J., Coles, R., Almqvist, E., Biancalana, V., Cassiman, J.-J., Chotai, K., Connarty, M., Craufurd, D., Curtis, A. et al. (1996) Phenotypic characterisation of individuals with 30–40 CAG repeats in the Huntington's disease (HD) gene reveals HD cases with 36 repeats and apparently normal elderly individuals with 36–39 repeats. *Am. J. Hum. Genet.* **58**, 16–22

10. Trottier, Y., Devys, D., Imbert, G., Saudou, F., An, I., Lutz, Y., Weber, C., Agid, Y., Hirsch, E.C. & Mandel, J.L. (1995) Cellular localization of the Huntington's disease protein and discrimination of the normal and mutated form. *Nat. Genet.* **10**, 104–110

11. White, J.K., Auerbach, W., Duyao, M.P., Vonsattel, J.-P., Gusella, J.F., Joyner, A.L. & MacDonald, M.E. (1997) Huntingtin is required for neurogenesis and is not impaired by the Huntington's disease CAG expansion. *Nat. Genet.* **17**, 404–410

12. David, G., Abbas, N., Stevanin, G., Dürr, A., Yvert, G., Cancel, G., Weber, C., Imbert, G., Saudou, F., Antoniou, E. et al. (1997) Cloning of the *SCA7* gene reveals a highly unstable CAG repeat expansion. *Nat. Genet.* **17**, 65–70

12a. Ross, C.A. (1995) When more is less: pathogenesis of glutamine repeat neurodegenerative diseases. *Neuron* **15**, 493–496

13. Zoghbi, H.Y. (1997) CAG repeats in SCA6: anticipating new clues. *Neurology* **49**, 1–5

14. Trottier, Y., Lutz, Y., Stevanin, G., Imbert, G., Devys, D., Cancel, G., Saudou, F., Weber, C., David, G., Tora, L. et al. (1995) Polyglutamine expansion as a pathological epitope in Huntington's disease and four dominant cerebellar ataxias. *Nature (London)* **378**, 403–406

15. Mangiarini, L., Sathasivam, K., Seller, M., Cozens, B., Harper, A., Hetherington, C., Lawton, M., Trottier, Y., Lehrach, H., Davies, S.W. & Bates, G.P. (1996) Exon 1 of the *HD* gene with an expanded CAG repeat is sufficient to cause a progressive neurological phenotype in transgenic mice. *Cell* **87**, 493–506

16. Davies, S.W., Turmaine, M., Cozens, B.A., Difiglia, M., Sharp, A.H., Ross, C.A., Scherzinger, E., Wanker, E.E., Mangiarini, L. & Bates, G.P. (1997) Formation of neuronal intranuclear inclusions underlies the neurological dysfunction in mice transgenic for the HD mutation. *Cell* **90**, 537–548

17. Skinner, P.J., Koshy, B., Cummings, C.J., Klement, I.A., Helin, K., Servadio, A., Zoghbi, H.Y. & Orr, H.T. (1997) SCA1 pathogenesis involves alterations in nuclear matrix associated structures. *Nature (London)* **389**, 971–974

18. Becher, M.W., Kotzuk, J.A., Sharp, A.H., Davies, S.W., Bates, G.P., Price, D.L. & Ross, C.A. (1998) Intranuclear neuronal inclusions in Huntington's Disease and dentatorubral and pallidoluysian atrophy: correlation between the density of inclusions and *IT15* CAG triplet repeat length. *Neurobiol. Dis.* **4**, 387–397

19. Difiglia, M., Sapp, E., Chase, K.O., Davies, S.W., Bates, G.P., Vonsattel, J.P. & Aronin, N. (1997) Aggregation of Huntingtin in neuronal intranuclear inclusions and dystrophic neurites in brain. *Science* **277**, 1990–1993

20. Paulson, H.L., Perez, M.K., Trottier, Y., Trojanowski, J.Q., Subramony, S.H., Das, S.S., Vig, P., Mandel, J.L., Fischbeck, K.H. & Pittman, R.N. (1997) Intranuclear inclusions of expanded polyglutamine protein in spinocerebellar ataxia type 3. *Neuron* **19**, 333–344

21. Igarashi, S., Koide, R., Shimohata, T., Yamada, M., Hayashi, Y., Takano, H., Date, H., Oyake, M., Sato, T., Sato, A. et al. (1998) Suppression of aggregate formation and apoptosis by transglutaminase inhibitors in cells expressing truncated DRPLA protein with an expanded polyglutamine stretch. *Nature Genet.* **18**, 111–117

22. Clark, H.B., Burright, E.N., Yunis, W.S., Larson, S., Wilcox, C., Hartman, B., Matilla, A., Zoghbi, H.Y. & Orr, H.T. (1997) Purkinje cell expression of a mutant allele of SCA1 in transgenic mice leads to disparate effects on motor behaviors followed by a progressive cerebellar dysfunction and histological alterations. *J. Neurosci.* **17**, 7385–7395

23. Matilla, A., Koshy, B., Cummings, C.J., Isobe, T., Orr, H.T. & Zoghbi, H.Y. (1997) The cerebellar leucine rich acidic nuclear protein (LANP) interacts with ataxin-1. *Nature (London)* **389**, 974–978

24. Li, X.J., Sharp, A.H., Li, S.H., Dawson, T.M., Snyder, S.H. & Ross, C.A. (1996) Huntingtin associated protein (HAP1): discrete neuronal localizations in the brain resemble those of neuronal nitric oxide synthase. *Proc. Natl. Acad. Sci. U.S.A.* **93**, 4839–4844

25. Kalchman, M.A., Koide, H.B., McCutcheon, K., Graham, R.K., Nichol, K., Nishiyama, K., Kazemi-Esfarjani, P., Lynn, F.C., Wellington, C., Metzler, M. et al. (1997) HIP1, a human homologue of *S. cerevisiae* Sla2p, interacts with membrane-associated huntingtin in the brain. *Nat. Genet,* **16,** 44–53

26. Ordway, J.M., Tallaksen-Greene, S., Gutekunst, C.A., Bernstein, E.M., Cearley, J.A., Wiener, H.W., Dure, L.S., Lindsey, R., Hersch, S.M., Jope, R.S. et al. (1997) Ectopically expressed CAG repeats cause intranuclear inclusions and a progressive late onset neurological phenotype in the mouse. *Cell* **91,** 753–763

27. Kosinski, C.M., Cha, J.H., Young, A.B., Persichetti, F., MacDonald, M., Gusella, J.F., Penney, Jr., J.B. & Standaert, D.G. (1997) Huntingtin immunoreactivity in the rat neostriatum: differential accumulation in projection and interneurons. *Exp. Neurol.* **144,** 239–247

28. Jarrett, J.T. & Lansbury, P.T.J. (1993) Seeding 'one-dimensional crystallization' of amyloid: a pathogenic mechanism in Alzheimer's disease and Scrapie? *Cell* **73,** 1055–1058

29. Goldberg, Y.P., Nicholson, D.W., Rasper, D.M., Kalchman, M.A., Koide, H.B., Graham, R.K., Bromm, M., Kazemi-Esfarjani, P., Thornberry, N.A., Vaillancourt, J.P. & Hayden, M.R. (1996) Cleavage of huntingtin by apopain, a proapoptotic cysteine protease, is modulated by the polyglutamine tract. *Nat. Genet.* **13,** 442–449

30. Scherzinger, E., Lurz, R., Turmaine, M., Mangiarini, L., Hollenbach, B., Hasenbank, R., Bates, G.P., Davies, S.W., Lehrach, H. & Wanker, E.E. (1997) Huntingtin-encoded polyglutamine expansions form amyloid-like protein aggregates *in vitro* and *in vivo. Cell* **90,** 549–558

31. Stott, K., Blackburn, J.M., Butler, P.J. & Perutz, M. (1995) Incorporation of glutamine repeats makes proteins oligomerize: implications for neurodegenerative diseases. *Proc. Natl. Acad. Sci. U.S.A.* **92,** 6509–6513

32. Ikeda, H., Yamaguchi, M., Sugai, S., Aze, Y., Narumiya, S. & Kakizuka, A. (1996) Expanded polyglutamine in the Machado–Joseph disease protein induces cell death in vitro and in vivo. *Nature Genet.* **13,** 196–202

33. Lunkes, A. & Mandel, J.L. (1998) A cellular model that recapitulates major pathogenic steps of Huntington's disease. *Hum. Mol. Genet.* **7,** 1355–1361

34. Kahlem, P., Green, H. & Djian, P. (1998) Transglutaminase action imitates Huntington's disease: selective polymerisation of huntingtin containing expanded polyglutamine. *Mol. Cell* **1,** 595–601

13

The matter of mind: molecular control of memory

Emily P. Huang and Charles F. Stevens[1]

Howard Hughes Medical Institute and Molecular Neurobiology Laboratory, The Salk Institute, 10010 N. Torrey Pines Rd., La Jolla, CA 92037, U.S.A.

Introduction

Until recently, psychologists and neurologists were the primary 'movers and shakers' in understanding how the mammalian brain holds memories. Research performed by these scientists revealed the types of memory that exist, the timing of memory formation, and the brain regions that are involved [1]. Ultimately, however, biologists aim to elucidate memory down to its molecular mechanisms — a daunting prospect, given that we have only the barest understanding of how memory works at the cellular level. Nonetheless, some key insights and the development of powerful molecular techniques have allowed us to dissect some of the biochemical processes involved in memory formation. In particular, this essay will discuss the hypothesis that memories are stored as a result of changes in synaptic connections between neurons, and will show how this idea has focused the attention of biologists on a mechanism of synaptic strengthening called long-term potentiation (LTP). We will also describe how researchers are using gene knock-out technology, which enables the creation of mice with specific gene mutations, to investigate whether LTP plays a role in behavioural memory.

As an alternative approach to studying the molecular basis of memory, biologists are also attacking simpler systems in which the link between molecules and behaviour is more directly accessible. Such systems include sensitization of gill withdrawal in the sea slug *Aplysia* and the odour-avoidance response of the fruit fly *Drosophila melanogaster*. These systems make up in

[1]*To whom correspondence should be addressed.*

tractability what they lack in similitude to mammalian memory, although there is reason to believe that memory systems in different organisms have some common molecular foundation. For instance, studies of these lower animals have shed light on the general problem of how memories are consolidated for long-term retention, a process that involves protein synthesis and the activation of gene transcription factors.

Synaptic basis of memory

The means by which brain neurons store and retrieve memories is a puzzle that biologists are still far from solving. A century ago, the neuroanatomist Ramon y Cajal proposed that the connections between neurons (i.e. their synapses) are altered in some manner when memories are formed [1]. Later, Donald Hebb extended this idea into a hypothetical rule of learning that would prove fruitful for scientists wishing to study the cellular basis of memory [2]. Hebb proposed that the process of learning involves intense or repeated activity at particular synapses, which become stronger as a result of the activity. This strengthening of synapses that are active during the learning process in turn forms the 'memory' of what has been learned.

For example, in Figure 1, neurons A and B both synapse on to neuron C. Under Hebb's proposal, if presynaptic neuron A were to fire and repeatedly activate postsynaptic neuron C, the synapse between these neurons would get stronger (that is, A will activate C more effectively). The A-to-C connection is

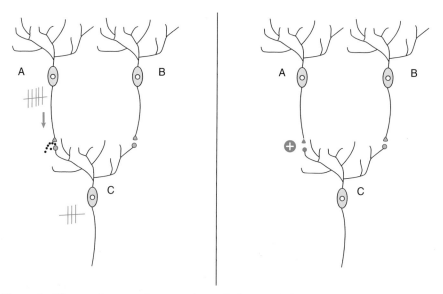

Figure 1. Changes in synaptic strength result from activity
Left: two neurons (A and B) synapse on to a third neuron (C). Repetitive activity in neuron A causes synaptic stimulation of neuron C. Right: the synapse between A and C is strengthened as a result of the activity in these neurons.

thus 'remembered', and the formation of specific patterns of such strengthened synaptic connections is a possible means of making memories. An important feature of this proposal is that a synapse will only strengthen if there is simultaneous activity in the presynaptic and postsynaptic neurons. Thus the synapse between B and C does not strengthen, because only C is active. Whether or not this Hebbian concept of memory storage is correct in all details, biologists have absorbed the general principle that memory depends on the modification of neuronal synapses, or *synaptic plasticity*. Those interested in the cellular basis of memory have thus focused their sights on processes by which synaptic connections are strengthened (or weakened) and the molecular mechanisms underlying these processes.

LTP

In 1973, T.V.P. Bliss and T. Lømo [3] made a pivotal discovery while examining a population of glutamate-releasing (glutamatergic) synapses in the hippocampus of the rabbit brain. Bliss and Lømo measured synaptic transmission in these synapses by stimulating the presynaptic neurons with single shocks and recording the postsynaptic response. In general, the level of synaptic transmission remained stable over time, but when they applied a short train of shocks at high frequency (100 Hz) to the synapses, the level of synaptic transmission rose sharply and remained elevated for hours afterwards (Figure 2). Bliss and Lømo had found a biological correlate to Hebb's rule — a strengthening of synapses in response to intense activity. They dubbed this phenomenon long-

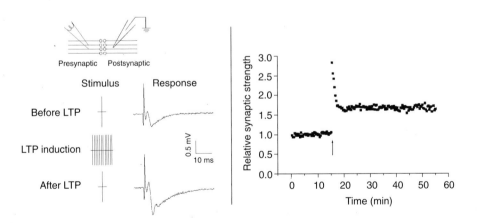

Figure 2. LTP in hippocampal synapses

Left: to measure synaptic transmission in a population of synapses, a stimulus electrode is placed among presynaptic axons and a recording electrode among the postsynaptic dendrites. Before and after LTP induction, single shocks are applied to the synapses and the postsynaptic response is measured. To induce LTP, a train of stimuli are applied at high frequency. Right: synaptic transmission in hippocampal synapses is plotted against time; the arrow indicates the time of LTP induction. Synaptic strength increases significantly after LTP. Data provided by D. Misner.

lasting potentiation, now known as LTP. More recently, physiologists have discovered an opposite partner to LTP, called long-term depression, which is a weakening of synaptic transmission in response to more moderate activity than that which induces LTP.

Excitement over LTP as a cellular mechanism for memory has remained high over the last two decades, for several reasons. First, the phenomenon, as discussed below, displays classically Hebbian properties that many scientists believe characterize the mechanisms underlying memory. Secondly, LTP is strongly expressed in the hippocampus, a cortical brain structure involved in the formation of declarative memory (i.e. the memory for factual information). Hopeful researchers thus envision LTP as the mechanism by which hippocampal neurons participate in memory formation. While time has neither proved nor disproved this hypothesis and the precise link between LTP and mammalian memory remains uncertain (as we will discuss later), the data so far are encouraging enough to merit a discussion of the molecular basis of LTP.

As we have noted, LTP is induced by strong or high-frequency stimulation of glutamatergic synapses. In these synapses, glutamate released from presynaptic terminals binds to glutamate receptor channels on the postsynaptic membrane, which open and allow cations to pass into the postsynaptic neuron. At least two types of glutamate receptor channels are present at cortical glutamatergic synapses: the N-methyl-D-aspartate (NMDA) receptor and the α-amino-3-hydroxy-5-methylisoxazolepropionate (AMPA) receptor. The properties of these glutamate receptors differ in several important respects [4]. For instance, AMPA receptors have fast kinetics and are permeant to Na^+ and K^+ ions, while NMDA receptors are slow-acting and also pass Ca^{2+} ions. In addition, NMDA receptor channels are normally blocked by Mg^{2+} ions, but this blockade is relieved if stimulus conditions depolarize the postsynaptic membrane, ejecting the Mg^{2+} from the channel pore.

When glutamate is released at the synapse, it binds to both AMPA and NMDA receptors. With a mild stimulus, the AMPA receptors allow a transient electrical signal in the form of Na^+ ions to pass into the postsynaptic neuron, but the NMDA receptors are blocked. With a strong stimulus, on the other hand, the AMPA receptors allow a large enough influx of cations to depolarize the postsynaptic membrane, unblocking the NMDA receptors and allowing them to pass Ca^{2+} ions into the postsynaptic neuron. This Ca^{2+} influx in turn triggers molecular cascades leading to the increase in synaptic strength called LTP (Figure 3). A battery of experiments confirm the dependence of LTP on NMDA receptor activity [5]; in particular, application of the NMDA receptor antagonist amino-5-phosphonopentanoate (AP5) during the high-frequency induction stimulus completely blocks LTP.

In summary, the NMDA receptor is the gateway through which a critical message (Ca^{2+}) must pass in order to turn on LTP. To open this gateway, two keys must be turned coincidentally: the presynaptic terminal must be active and releasing transmitter, and the postsynaptic cell must be active and depolar-

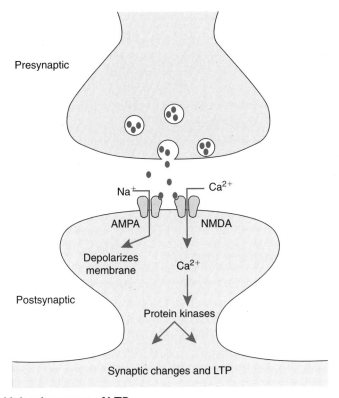

Figure 3. Molecular events of LTP
Released glutamate binds to AMPA receptors, depolarizing the postsynaptic membrane. The NMDA receptor detects simultaneous glutamate binding and membrane depolarization, and allows Ca^{2+} into the postsynaptic neuron. This Ca^{2+} influx triggers the synaptic changes associated with LTP.

ized. For this reason the NMDA receptor is often called a *coincidence detector*. The NMDA receptor thus confers the Hebbian property of associativity on LTP, allowing it to occur if, and only if, the presynaptic and postsynaptic cells are simultaneously active.

How this NMDA receptor-mediated postsynaptic Ca^{2+} influx subsequently leads to an increase in synaptic strength is not yet clear, despite strenuous research efforts [6,7]. In particular, it is not agreed whether the increase in synaptic strength seen in LTP is the result of an increase in the release of transmitter, an increase in the response to transmitter, the appearance of new synapses, or some combination of these effects. Nonetheless, scientists have tested a host of molecules for their involvement in the expression of LTP, the most prominent of which are the protein kinases. Early experiments [5] showed that non-specific kinase inhibitors such as H-7 and sphingosine block the expression of LTP. Furthermore, several protein kinases are highly expressed in the hippocampus, including Ca^{2+}/calmodulin-dependent kinase (CaMKII), protein kinase C and the tyrosine kinase Fyn. Protein kinase C and CaMKII are activated directly or indirectly by exposure to Ca^{2+}, providing a link to the NMDA receptor-mediated induction events of LTP. Further evidence that kinases are involved in LTP will be discussed in the following section.

Potential targets for the activity of these kinases depend on whether LTP involves an increase in transmitter release or an increase in receptor sensitivity (or both). Two proposals are that kinases up-regulate postsynaptic AMPA receptor activity [8] and modulate synapsin, a transmitter vesicle-associated protein that regulates transmitter release [9]. Besides these proteins, other molecules have been shown to be involved in LTP, including G-protein-coupled glutamate receptors [10], neurotrophins [11] and nitric oxide (see Chapter 7 in the present volume). Ultimately, however, we must clarify the locus of LTP expression before we fully understand how these molecules of LTP work together to strengthen synaptic transmission.

Mutant mice, memory and LTP

Whether LTP truly underlies memory formation is a contentious question, despite the research effort expended on this phenomenon. Because we know so little about the neuronal circuitry of memory, it is difficult to show a direct relationship between LTP *in vivo* and memory, as tested by animal behaviour. One of the first significant attempts to demonstrate such a relationship focused on the dependence of LTP on NMDA receptor activity. In a classic experiment, R.G.M. Morris et al. [12] infused the hippocampus of rats with the NMDA receptor antagonist AP5 and tested the rats for their ability to form spatial memories. In the Morris watermaze task, a widely used test for rat spatial memory, rats are placed in a pool of opaque water with a submerged platform, which they locate in the course of swimming around the pool. Using visual cues in the environment, the rats learn the spatial location of the platform and afterwards swim directly to it, regardless of where they are placed in the pool.

Rats infused with AP5 in the hippocampus are severely impaired in the watermaze task, implicating NMDA receptor activity and, by extension, LTP in rat spatial memory. Unfortunately, such behavioural/pharmacological studies are questionable because of the non-specific effects (such as sensorimotor disturbances) that could be mediated by the blockade of NMDA receptors. Similarly, using gene knock-out technology to create mice lacking a functional NMDA receptor is not useful for memory research, because such mutants die in the neonatal period. Instead, molecular biologists have focused on creating mice lacking molecules downstream of the NMDA receptor in the LTP cascade, such as protein kinases. Mice with functionally inactive CaMKII and Fyn are severely impaired in spatial memory (as tested by watermaze learning) and LTP expression, which at least circumstantially supports the belief that LTP underlies memory [13–15].

These studies, however, bear a number of caveats. First, it is possible that mice developing without these kinases acquire non-specific developmental defects. Secondly, these experiments tend to ask separate questions: whether the kinase mutant mouse is defective in LTP and whether it is defective in memory. If there appears to be a defect in LTP *and* the mouse is memory-impaired, the experimenters tend to conclude that these phenotypes are cause and effect. This interpretation has obvious drawbacks. For instance, these kinases may be involved in phenomena other than LTP that underlie memory. Because we do not yet know precisely how hippocampal neurons perform their memory-related functions, nor indeed exactly what these functions are, it is difficult to design a more sophisticated approach, and such doubts will continue to dog molecular-based efforts to link LTP and memory.

Recent advances in gene knock-out technology have the potential to greatly refine the study of mouse memory mutants. These desirable changes allow the experimenter to control when the gene will be knocked out and in what cell types. In particular, Tsien et al. [16] used a new method of cell-restricted gene deletion [17] to create mice lacking the NMDA receptor in subregions of the forebrain, such as the hippocampus. This method utilizes the Cre/loxP recombination system, in which Cre, a bacteriophage recombinase enzyme, catalyses recombination between specific sequences called loxP sites. To create a cell-type-restricted knock-out using this system, one creates two mouse strains, each of which will be phenotypically normal (Figure 4). In one mouse strain, the gene for Cre is placed behind a cell-type-restricted promoter sequence; in another mouse strain, loxP sequences are placed around the gene one wishes to delete. When these mouse strains are crossed, Cre, which is expressed only where the promoter is active, catalyses the recombination of the loxP sequences, resulting in gene deletion in those cells.

Tsien et al. [16] placed the *Cre* transgene behind the αCaMKII promoter, restricting its expression to areas of the forebrain; in some of the mouse strains obtained this way, the expression of Cre was restricted to the hippocampus alone. They created another mouse line with loxP sequences flanking the NMDA receptor gene and then crossed these lines, producing mice lacking the NMDA receptor in hippocampal neurons. Because the αCaMKII promoter is chiefly driven postnatally, this method also minimized potential developmental problems in the knock-out. Tsien et al. examined the affected synapses of these mice and found, as expected, that LTP expression was impaired. Furthermore, spatial memory in these mice, as tested by the watermaze task, was also impaired. Since the molecular defect in this study was limited to a small region and the causal link between the NMDA receptor and LTP is well established, this result provides the best evidence derived so far from molecular techniques that LTP (or a closely related phenomenon) is involved in the formation of spatial memory.

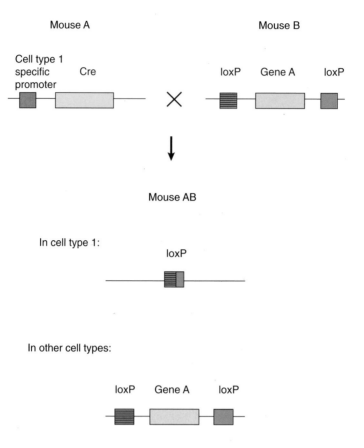

Figure 4. Gene deletion in specific cell types
Mouse A, which expresses the recombinase enzyme Cre in cell type 1, is crossed with Mouse B, which has loxP sequences flanking Gene A. In their progeny, Cre recombination of the loxP sequences leads to deletion of Gene A from cell type 1.

Drosophila memory and cAMP

As some investigators studied synaptic plasticity in mammalian brains, others were doing work in simpler systems where one might reasonably hope to make direct connections between molecules and behavioural memory. *Drosophila*, while not famous for brain power, is nonetheless useful in the search for memory molecules because it is relatively easy to screen for genes that affect fly behaviour. In particular, investigators have focused on a simple form of unconscious memory called classical conditioning. In the most frequently used behavioural paradigm, flies are given electric shocks at the same time as they are exposed to particular odours. As one might expect, the flies learn to avoid the specific odours that have been associated with the shocks, in effect forming olfactory memories.

Flies with single-gene mutations are generated by a number of methods (chemical mutagenesis, transposable element insertions) and can be screened for behavioural defects, such as an inability to remember odour–shock associations. Any fly strain that shows an inherited inability to perform this memory task is a potential memory mutant (excluding those that cannot perform the task because of sensory or motor defects, which can be tested independently). The best known of such mutant strains are *dunce* and *rutabaga*; these mutants are capable of learning odour–shock associations, but lose the resulting memories very quickly. Based on analogies with classical conditioning in the sea slug *Aplysia*, fly geneticists explored the possibility that these mutants were defective in cAMP metabolism [18,19]. Eventually, the *dunce* gene was cloned and proven to encode cAMP phosphodiesterase. Similarly, the *rutabaga* gene encodes a form of adenylate cyclase.

How do defects in cAMP metabolism relate to the hypothesis that memory is formed via changes in synaptic connectivity? In *Aplysia*, it has been shown that classical conditioning leads to a rise in cAMP in synapses that subserve the learned behaviour. This rise in cAMP activates cAMP-dependent protein kinase (PKA), which in turn phosphorylates target proteins, leading to increased transmitter release and synaptic growth [20]. If fly memory works analogously, defects in the cAMP pathway, such as those seen in *dunce* and *rutabaga* mutants, might impair the fly's ability to modify its synapses during memory formation. Studies have shown that the plasticity of neuromuscular synapses in *dunce* and *rutabaga* mutants is different from that in normal flies [21], but the physiological properties of synapses in the fly brain, particularly those that may subserve the formation of olfactory memories, remain largely unexplored.

Long-term memory in *Drosophila*

The defects in *dunce* and *rutabaga* disrupt the early stages of memory formation and are thus uninformative about the later stages. Of particular interest is the process by which acquired memories are consolidated for long-term storage. For instance, after learning an odour-avoidance task, fruit flies are capable of remembering which odours to avoid for up to 1 week, for a significant fraction of their lifetime. Work from T. Tully's laboratory [22] demonstrated that whether or not a fly acquires long-term olfactory memories depends on the manner in which it is taught the odour–shock association. If it is trained in a series of sessions with no break in between (massed training), its memory of the association completely disappears within 4 days. However, if it is trained in the same number of sessions, but at 15 min intervals (spaced training), the memory remains robust for at least 7 days.

For some time, scientists have recognized that consolidating memories for future access might require mechanisms beyond biochemical cascades at synapses. In particular, it was proposed that new gene transcription and protein synthesis must occur in order to make the long-term changes in memory. In support of this idea, Tully's group found that, when flies were fed the protein synthesis inhibitor cycloheximide (mixed temptingly with glucose) just before spaced training, their memory of the odour–shock association decayed precipitously, much like what is seen after massed training. This discovery implied that the formation of long-term memory in flies was indeed dependent on the synthesis of new proteins, which in turn implied the involvement of new gene transcription and at least one inducible transcription factor. Since cAMP was the chief suspect in fly memory, Tully's group postulated that the transcription factor in question might be cAMP-inducible. Again, the foundation for this speculation was laid by work in the *Aplysia* memory system, where it had been found that cAMP-response-element-binding protein (CREB), a transcription factor activated by elevations in cAMP, plays a critical role in the formation of long-term memory [20]. CREB is a member of the basic region/leucine-zipper transcription factor family, binds to specific enhancer sequences called CREs (cAMP response elements), and is activated by PKA-mediated phosphorylation of a single serine residue.

To test the role of CREBs in fly memory, the Tully group transgenically expressed a dominant-negative form of CREB called *d*CREB2-b (an isoform that acts as a repressor of CREB-mediated transcription) under a heat-shock promoter [23]. When they transiently induced expression of this CREB repressor during spaced training, the transgenic flies behaved like those that had been fed with cycloheximide: they were unable to form long-term olfactory memories. On the other hand, flies transgenically expressing an activator isoform (*d*CREB2-a) formed long-lasting olfactory memories after only one training session [24]; in these flies, multiple training sessions are no longer necessary to consolidate memories.

Together, these studies demonstrate the integral role of CREB-mediated gene transcription in fly long-term memory, and imply that flies normally require spaced training to build up sufficient CREB levels to form such memories. The genes activated by CREB during long-term memory consolidation are largely unknown. In *Aplysia*, experiments have shown that CREB induces the transcription of immediate-early genes, some of which are themselves transcription factors, but parallel work remains to be done in flies.

CREB and mammalian memory

Compelling and attractive as these results from *Drosophila* and *Aplysia* might be, are they relevant to mammalian memory? Recent studies have shown that mice with targeted mutations in CREB are significantly impaired in their ability to perform the Morris watermaze task, implying that CREB-mediated transcription may indeed play a role in mammalian memory. Similarly, mice transgenically expressing an inhibitory form of a PKA subunit specifically in the forebrain are also impaired in spatial memory. In addition, these CREB mutants and PKA transgenics display forms of hippocampal LTP that decay more rapidly than in the wild-type, indicating that CREB activation may play a role in maintaining LTP expression over long periods of time.

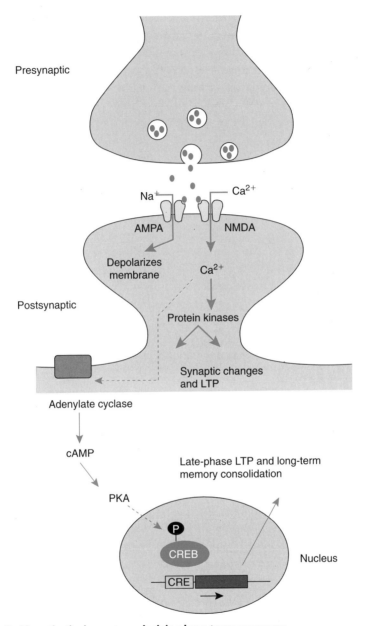

Figure 5. Hypothetical events underlying long-term memory
The Ca^{2+} influx from LTP induction stimulates the production of cAMP, which activates PKA. PKA directly or indirectly activates CREB, causing transcription of genes with CRE enhancer elements. The resulting protein synthesis leads to late-phase LTP and long-term memory consolidation.

These results have led some scientists to formulate a hypothetical picture of the molecular events underlying long-term memory formation in the hippocampus, which is depicted in Figure 5. In this scheme, the molecular events of LTP lead to a synaptic rise in cAMP through the activation of adenylate cyclase. The rise in cAMP and subsequent activation of PKA lead either direct-ly or indirectly to CREB activation, turning on gene transcription and ulti-mately leading to the expression of a late phase of LTP that is biochemically distinct from earlier LTP phases. Presumably, the expression of late-phase LTP underlies the consolidation of long-term memory.

Perspectives

Recent improvements in knock-out technology promise to bring us much new information about the molecular cascades underlying memory. In particular, methods for creating temporally controlled [25] (as well as cell-restricted) knock-outs are being developed and will improve the use of mutants to study behavioural and cognitive functions. Nonetheless, even such finely tuned mol-ecular studies of memory will be difficult to interpret in the absence of detailed knowledge about the neuronal interactions that create memories. In dissecting such a complex function, it will be necessary to integrate studies of molecular interactions with experiments on neuronal networks and behaviour [26]. One possible option is to study simpler mammalian memory systems in which the underlying circuitry is better understood than in the hippocampus. Fear condi-tioning, i.e. the association of a previously neutral stimulus with fearful emo-tion, is a form of learning that depends on characterized neural pathways in the amygdala. Recent studies have demonstrated that fear conditioning induces LTP in amygdala synapses thought to underlie the stimulus–fear association [27,28]. If the implications of these experiments are confirmed, the amygdala may prove to be an attractive new system in which to study the molecular basis of memory.

Summary

- *A widely accepted hypothesis suggests that changes in synaptic strength underlie the formation of memories in the brain.*
- *LTP is a mechanism of synaptic strengthening. Induction of LTP depends on NMDA receptor activation, and its expression depends in part on protein kinase activity.*
- *Studies of knock-out mice suggest that LTP is critical for hippocampus-based memory.*
- *Genetic studies in Drosophila implicate cAMP metabolism in classical conditioning, a form of unconscious memory.*
- *Consolidating memories for long-term retention depends on the cAMP-inducible transcription factor CREB.*

References

1. Squire, L. (1987) *Memory and Brain*, Oxford University Press, New York
2. Hebb, D.O. (1949) *The Organization of Behavior*, John Wiley & Sons, New York
3. Bliss, T.V.P. & Lømo, T. (1973) Long-lasting potentiation of synaptic transmission in the dentate area of the anaesthetized rabbit following stimulation of the perforant path. *J. Physiol. (London)* **232**, 331–356
4. Hollmann, M. & Heinemann, S. (1994) Cloned glutamate receptors. *Annu. Rev. Neurosci.* **17**, 31–108
5. Madison, D.V., Malenka, R.C. & Nicoll, R.N. (1991) Mechanisms underlying long-term potentiation of synaptic transmission. *Annu. Rev. Neurosci.* **14**, 379–397
6. Stevens, C.F. & Wang, Y. (1994) Changes in reliability of synaptic function as a mechanism for plasticity. *Nature (London)* **371**, 704–707
7. Liao, D., Hessler, N.A. & Malinow, R. (1995) Activation of postsynaptically silent synapses during pairing-induced LTP in CA1 region of hippocampal slice. *Nature (London)* **375**, 400–404
8. Barria, A., Muller, D., Derkach, V., Griffith, L.C. & Soderling, T.R. (1997) Regulatory phosphorylation of AMPA-type glutamate receptors by CaM-KII during long-term potentiation. *Science* **276**, 2042–2045
9. Nayak, A.S., Moore, C.I. & Browning, M.D. (1996) Ca^{2+}/calmodulin-dependent protein kinase II phosphorylation of the presynaptic protein synapsin I is persistently increased during long-term potentiation. *Proc. Natl. Acad. Sci. U.S.A.* **93**, 15451–15456
10. Lu, Y.-M., Jia, Z., Janus, C., Henderson, J.T., Gerlai, R., Wotjowicz, J.M. & Roder, J.C. (1997) Mice lacking metabotropic glutamate receptor 5 show impaired learning and reduced CA1 long-term potentiation (LTP) but normal CA3 LTP. *J. Neurosci.* **17**, 5196–5205
11. Patterson, S.L., Abel, T., Deuel, T.A.S., Martin, K.C., Rose, J.C. & Kandel, E.R. (1996) Recombinant BDNF rescues deficits in basal synaptic transmission and hippocampal LTP in BDNF knockout mice. *Neuron* **16**, 1137–1145
12. Morris, R.G.M., Andersen, E., Lynch, G. & Baudry, M. (1986) Selective impairment of learning and blockade of long-term potentiation by an *N*-methyl-D-aspartate receptor antagonist, AP5. *Nature (London)* **319**, 774–776
13. Silva, A.J., Stevens, C.F., Tonegawa, S. & Wang, Y. (1992) Deficient hippocampal long-term potentiation in α-calcium-calmodulin kinase II mutant mice. *Science* **257**, 201–206
14. Silva, A.J., Paylor, R., Wehner, J.M. & Tonegawa, S. (1992) Impaired spatial learning in α-calcium-calmodulin kinase II mutant mice. *Science* **257**, 206–211
15. Grant, S.G.N., O'Dell, T.J., Karl, K.A., Stein, P.L., Soriano, P. & Kandel, E.R. (1992) Impaired long-term potentiation, spatial learning, and hippocampal development in fyn mutant mice. *Science* **258**, 1903–1910
16. Tsien, J.Z., Huerta, P.T. & Tonegawa, S. (1996) The essential role of hippocampal CA1 NMDA receptor-dependent synaptic plasticity in spatial memory. *Cell* **87**, 1327–1338
17. Gu, M., Marth, J.D., Orban, P.C., Mossmann, H. & Ragewsky, K. (1994) Deletion of a DNA polymerase β gene segment in T cells using cell type-specific gene targeting. *Science* **265**, 103–106
18. Levin, L.R., Han, P.-L., Hwang, P.M., Feinstein, P.G., Davis, R.L. & Reed, R.R. (1992) The *Drosophila* learning and memory gene *rutabaga* encodes a Ca^{2+}/calmodulin-responsive adenylyl cyclase. *Cell* **68**, 479–489
19. Byers, D., Davis, R.L. & Kiger, J.A. (1981) Defect in cyclic AMP phosphodiesterase due to the *dunce* mutation of learning in *Drosophila melanogaster*. *Nature (London)* **289**, 79–81
20. Bailey, C.H., Bartsch, D. & Kandel, E.R. (1996) Toward a molecular definition of long-term memory storage. *Proc. Natl. Acac. Sci. U.S.A.* **93**, 13445–13452
21. Davis, R.L. (1993) Mushroom bodies and *Drosophila* learning. *Neuron* **11**, 1–14
22. Tully, T., Preat, T., Boynton, S.C. & Del Vecchio, M. (1994) Genetic dissection of consolidated memory in *Drosophila*. *Cell* **79**, 35–47

23. Yin, J.C.P., Wallach, J.S., Del Vecchio, M., Wilder, E.L., Zhou, H., Quinn, W.G. & Tully, T. (1994) Induction of a dominant negative CREB transgene specifically blocks long-term memory in *Drosophila*. *Cell* **79**, 49–58

24. Yin, J.C.P., Del Vecchio, M., Zhou, H. & Tully, T. (1995) CREB as a memory modulator: induced expression of a dCREB2 activator isoform enhances long-term memory in *Drosophila*. *Cell* **81**, 107–115

25. Mayford, M., Bach, M.E., Huang, Y.-Y., Wang, L., Hawkins, R.D. & Kandel, E.R. (1996) Control of memory formation through regulated expression of a CaMKII transgene. *Science* **274**, 1678–1683

26. McHugh, T.J., Blum, K.I., Tsien, J.Z., Tonegawa, S. & Wilson, M.A. (1996) Impaired hippocampal representation of space in CA1-specific NMDAR1 knockout mice. *Cell* **87**, 1339–1349

27. Rogan, M.T., Staubli, U.V. & LeDoux, J.E. (1997) Fear conditioning induces associative long-term potentiation in the amygdala. *Nature (London)* **390**, 604–607

28. McKernan, M.G. & Shinnick-Gallagher, P. (1997) Fear conditioning induces a lasting potentiation of synaptic currents *in vitro*. *Nature (London)* **390**, 607–611

14

Future developments

Susan Greenfield

Department of Pharmacology, University of Oxford, Mansfield Road, Oxford, Oxon OX1 3QT, U.K.

Introduction

Any predictions on future developments in brain research are only worthwhile and interesting if, as for any branch of science, they challenge current assumptions at a really fundamental level. In the case of the brain, the most accepted framework for exploring function and functioning is, arguably, synaptic transmission: substances, 'transmitters', are released from axonal nerve terminals in a quantal and local fashion to activate specific, closely apposed targets to cause either 'inhibition' or 'excitation', namely the suppression or generation respectively of action potentials. Perhaps one of the most sweeping developments in our understanding in the future will be to question the seeming monopoly of this familiar sequence of events as a satisfactory means for accounting for the workings of the brain: over the last decade, it has become hard to explain an increasing number of observations in terms merely of a binary on/off scheme.

In this concluding chapter I shall consider some of the principal anomalies arising from current findings, specifically why: (a) there are many diverse transmitter substances; (b) transmitters are released from sites outside of the classical synapse; (c) some well-known transmitters have surprising, 'modulatory' actions; (d) synaptic mechanisms themselves have no obvious or direct one-to-one relationship with functions such as movement, mood and memory; and (e) it is difficult to extrapolate from drug-induced modification of synaptic mechanisms to the effects of those same drugs in influencing disorders such as depression, Parkinson's disease and schizophrenia.

*To whom correspondence should be addressed.

Why are there so many different neurotransmitters?

The five established criteria for a transmitter are: (i) there should be evidence of its presence within the presynaptic cell; (ii) that it should exhibit calcium-dependent exocytosis; (iii) that it should act on a specific receptor site expressed on the surface of the receiving neuron; (iv) that it should be removed rapidly from the site of action; (v) that its action should be mimicked by presynaptic stimulation. The list of putative transmitters in the brain is growing all the time (Table 1), but some newcomers, such as the gaseous transmitters (see Chapter 7 in the present volume), do not comply with all the prerequisites, such as exocytosis or specific receptor-mediated action.

Another departure from the classical way of thinking has been triggered by observations of the distribution and behaviour of peptides (see Chapter 6) in the brain. It had previously been accepted that there was only ever one transmitter for one neuron. Soon it was realized, however, that each of the peptides was quite frequently stored with a classical transmitter (Table 2) [1]. Does such a situation then imply that peptides are not, after all, transmitters? An alternative interpretation has been offered by Hokfelt [1], who has suggested that the peptides and classical transmitters can be released either separately or together, according to the status of excitability of the cell. If the neuron is firing relatively slowly, then only small clear vesicles, containing transmitter alone, will be pressed into service at the synaptic cleft. However, as the discharge rate of the cell increases, then gradually an increasing number of dense-core vesicles will be recruited that contain both the transmitter and the peptide. As the discharge rate reaches the types of level reminiscent of pathologies, then it is the peptide release that will dominate. Hence the peptides appear to provide a mechanism for transforming a *quantitative* factor, cell firing, into a *qualitative* one, of relevance to the subsequent response of the target cell.

A further phenomenon, until recently regarded as heretical but now gaining credence, is that an enzyme can be released to contribute to neuronal signalling in a non-catalytic capacity. Although well known for its indispensable

Table 1. Neurotransmitters and candidate neurotransmitters: some examples

Type	Examples
Phenethylamines	Dopamine, noradrenaline, adrenaline
Indoleamines	5-Hydroxytryptamine (serotonin), tryptamine, melatonin
Cholinergics	Acetylcholine (ACh), choline
Neuropeptides	Enkephalins, cholecystokinin (CCK), thyrotropin-releasing hormone (TRH)
Amino acids	Glutamate, aspartate, glycine, γ-aminobutyric acid (GABA)
Nucleotides	Adenosine, ATP
Gases	Nitric oxide, carbon monoxide

Table 2.   Neuroactive peptides co-exist with other transmitters

Transmitter	Peptide	Location
γ-Aminobutyric acid (GABA)	Motilin	Cerebellum
	Cholecystokinin (CCK)	Cortical neurons
Acetylcholine (ACh)	Vasointestinal peptide (VIP)	Parasympathetic and cortical neurons
	Substance P	Pontine neurons
Noreprinephrine	Somatostatin	Sympathetic neurons
	Enkephalin	Sympathetic neurons
	Neuropeptide Y(NPY)	Medullary and pontine neurons
	Neurotensin	Locus ceruleus
Dopamine	CCK	Ventrotegmental neurons
	Neurotensin	Ventrotegmental neurons
Adrenaline	Neuropeptide Y(NPY)	Reticular neurons
	Neurotensin	Reticular neurons
5-Hydroxytryptamine (serotonin)	Substance P	Medullary raphe
	Thyrotropin-releasing hormone (TRH)	Medullary raphe
	Enkephalin	Medullary raphe
Vasopressin	CCK	Magnocellular hypothalamic neurons
	Dynorphin	Magnocellular hypothalamic neurons
Oxytocin	Enkephalin	Magnocellular hypothalamic neurons

catalytic action in terminating the action of acetylcholine (ACh), acetyl-cholinesterase is also found in brain regions with relatively little of its normal substrate ACh, or indeed its synthesizing enzyme choline acetyltransferase. Moreover, acetylcholinesterase also exists as a soluble form that is still catalytically active, but which, unlike the traditional membrane-bound form, can be released. Such release is not, of course, a requirement for cholinergic transmission, and suggests that the protein, or more specifically a non-catalytic site within it, has an alternative, non-enzymic action [2]. But again, could such an action of a protein be expected to compare with that of the smaller, and far less stable, classical transmitters?

No doubt the forthcoming years will herald the discovery of still further surprising transmitter-like molecules that strain the accepted concept of how transmitters behave. The diversity in size and type of molecule might well reflect different roles in signalling that do not fit easily into the current concept of 'inhibition' or 'excitation'. Just as important as the molecular identity of the substance in question will be the manner of its externalization from the neuron (see Chapter 3), and its subsequent fate in the extracellular space.

Why should neurotransmitters be released from outside of the classical synapse?

It is now known that neurons can send signals by means other than via the familiar axonal synapse, using all the classes of transmitter mentioned in the previous section. Noradrenaline and 5-hydroxytryptamine (serotonin) have been described in vesicles lying outside of the synaptic region, in neurons in the cerebral cortex, locus ceruleus and hippocampal formation [3]. Amino acids too may also participate in non-synaptic communication, for example [3] in neuron–glia interactions, where the importance of glia in the metabolism of glutamate and γ-aminobutyric acid (GABA) has long been recognized. Indeed, it is often overlooked that glutamate, a transmitter in its own right, is also a precursor molecule in the synthesis of GABA. Once freed from the tight spatial restriction of a classical synapse, it might be possible that these two substances, which most often have opposite effects on membrane potential (i.e. depolarization and hyperpolarization), could have antagonistic, and thus balancing, homoeostatic effects.

Indeed, the distribution of large, dense-core vesicles, which we saw in the preceding section to be the most typical storage site for peptides, is characteristically extrasynaptic, compared with the synaptic distribution of the more familiar and ubiquitous small vesicles which contain the smaller and more traditional transmitters. This differential localization would support the idea that peptides might be fulfilling a totally different type of signalling, compared with classical synaptic transmission, that entails a far less specific, and far more prolonged, action.

Sometimes the sites of release are not just outside the zone of the synapse, but in a totally different part of the neuron altogether, the dendrites. Dendritic release of transmitter has so far been best characterized for dopamine in the nigrostriatal pathway [4]. Until recently any investigation of, say, the dendritic release of dopamine within the substantia nigra was dogged by the fact that existing assay techniques had a time resolution of minutes and sampled over hundreds and thousands of cells. But now the use of fast-scan cyclic voltammetry (Figure 1) has enabled virtual on-line measurement and simultaneous identification, at least of oxidizable substances such as the amines, over sub-second periods such as 4 Hz [5,6]. It is now apparent that, even within the same cells, dendritic release within the substantia nigra is markedly different from the release of dopamine from axonal terminals in the striatum [5–7], and even from an adjacent midbrain area that also releases dopamine, the ventral tegmental area [5,6].

Similarly, the spatial resolution afforded by fast-scan cyclic voltammetry is now high enough to distinguish heterogeneous neuronal subpopulations within an erstwhile seemingly uniform dopaminergic cell population, such as that of the substantia nigra itself [8]. Moreover, compared with release from axon terminals, there are not only anatomical differences within the midbrain, such as lower numbers of synapses, but also quantitative differences in the actual uptake mechanisms controlling the extracellular availability of dopamine released from dendrites. These quantitative discrepancies lead to qualitative differences in signalling [5–7].

The time and space sensitivity for neurochemical detection is almost finally commensurate with that of electrophysiological measurements: over a few milliseconds from a single cell. Such intracellular recordings have revealed that dendrites not only release transmitters, but are also capable of generating autonomous or semi-autonomous potentials [8]. In the Purkinje cells of the cerebellum, for example, calcium conductances are operational in the dendrites independent of the state of excitability of the cell body; this scenario also occurs elsewhere in the brain, for example in neurons in the thalamus [9], the hippocampus [10] and the substantia nigra [11].

The generation of dendritic potentials and the dendritic release of transmitter might well occur independently of the generation of action potentials at the cell body, and thus both observations highlight the dendrites as a novel site for a non-classical system of neuronal communication [12]. For what types of function might autonomous dendritic signalling be valuable?

A further unexpected finding is that, in development, the mechanism of regulation of extracellular dopamine release from axons — prior to the formation of mature dopamine synapses — strongly resembles that from dendrites in the mature brain [13]. Hence dendritic release in the adult might be fulfilling, albeit at a more modest level, the kind of function that was more dominant and commonplace in many more types of neuronal population in the growing brain. Interestingly enough, there is a certain group of neurons in the brain

a   b

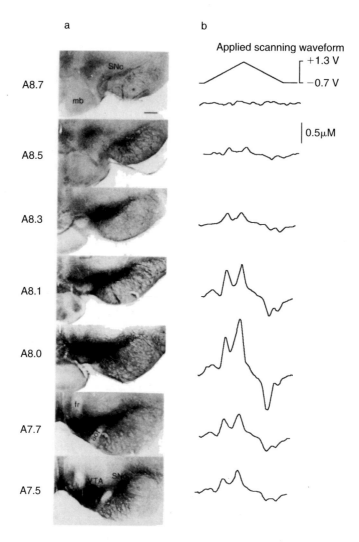

Figure 1. **Use of fast-scan cyclic voltammetry to monitor the release of dopamine from dendrites in the substantia nigra, with a high spatial and temporal resolution**
(a) Rostrocaudal variation in density of cells showing immunoreactivity for tyrosine hydroxylase. Abbreviations: mb, mamillary bodies; SNc, substantia nigra, pars compacta; fr, fasciculus retroflexus ; aot, accessory optic tract; VTA, ventral tegmental area. A8.7–A7.5 indicate the distance in mm from interaural line. (b) Rostrocaudal variation in stimulated external dopamine concentration along the rostrocaudal axis; $n = 10$–36 measurements from medial and lateral sites combined (number of cells per unit sample volume; grid sample volume 175 μm × 350 μm × 50 μm) in the same slices as in (a). $n = 6$–22 measurements at each slice co-ordinate. All data points are means ± S.E.M. Note the parallel variation in histochemical stain and electrochemical signal. Reproduced from [5] with permission. © 1997 American Physiological Society.

that retain some regenerative, developmental capacities into maturity [14]. This group of neurons (comprising the basal forebrain, ventral tegmental area, substantia nigra, hypothalamus, locus ceruleus, raphe nuclei and upper motor neurons) all have diffuse dendritic arbours with promiscuous, non-specific connectivity which would be particularly appropriate for the dendritic release of transmitter as the predominant form of signalling. It is possible, therefore, that within and across these cell groups there is a non-classical type of communication. This is ideally mediated by dendrites and is used most widely in development before precise synapses are established, and thus entails a more widespread, less localized type of signalling when compared with familiar synaptic transmission. But to appreciate the effects of these agents released so much more promiscuously into the extracellular space, we need to consider the potential target sites: both the relevant cells and their particular receptors (see Chapter 2).

How can familiar transmitters have unpredictable actions?

Extrasynaptic receptors have been located for G-protein-linked receptors employing peptides, monoamines and glutamate as neurotransmitters, as well as for ion-channel-gated receptors [3]. As might be predicted for a target site that will receive a relatively low concentration of diffused transmitter with a long latency, these extrajunctional receptors generally have a higher affinity than their counterparts in the synapse and are commonly metabotrobic [3]. For example, D_2-dopamine receptors on photoreceptors in the retina are several tens of micrometres away from any dopaminergic nerve terminal. Moreover, within the brain itself, D_2 receptors are selectively located at sites associated with the spines on dendrites of striatal neurons [15], namely on the very part of the cell most readily changed during more long-term development, where imaging techniques have now captured the retraction and expansion of spines over periods as brief as 30 min [16]. It is possible therefore, that D_2 receptors, so central in the pharmacology of schizophrenia (see Chapter 9), play a role other than in the fast, local transfer of information that is so characteristic of classical synaptic transmission.

A more established, but still non-classical, action of the familiar transmitter ACh is to block certain potassium channels mediating the 'm current', which in turn are only activated when certain cells, for example those in the hippocampus, are depolarized [7]. Hence ACh is ineffective unless released in conjunction with a depolarizing event, in which case the resultant action potentials will be generated more vigorously than otherwise.

A similar effect, i.e. a contingent modulation rather than a completely predictable triggering of action potentials, can also be achieved with ACh using an opposite mechanism, namely an opening of potassium channels. More specifically, in the thalamus ACh will open potassium channels with properties different from those mediating the m current. This opening of potassium channels leads to a hyperpolarization of the cell [17], which serves as the prerequisite condition for de-inactivation of the T-type calcium channel. Now, if depolarization occurs from this hyperpolarized level, the subsequent calcium influx will ensure a greater generation of action potentials. By a comparable hyperpolarizing action, GABA can fulfil a similar function in the hippocampus [18]. Conversely, it has been known for some 10 years that transmitters such as noradrenaline [19] and substance P [20] can have non-classical actions in enhancing 'modulating' calcium currents more directly. Hence the existence of channels other than those mediating the classical action potential, and indeed the fact that such conductances are frequently localized in dendrites, highlights the diversity and flexibility of classical transmitters acting in a more modulatory role, within the context of the geometry of the neuron and even certain time frames.

Not only can the voltage-dependent channels, such as the T-channel, act as a valuable mechanism for predisposing the neuron to certain responses conditional on certain contingencies, but a receptor with a built-in requirement for a specific pairing of otherwise independent events acts as an even more powerful calcium ionophore, i.e. the N-methyl-D-aspartate (NMDA) receptor complex (see Chapter 13). It should not be forgotten, however, that the efficacy of the NMDA receptor can be further modulated by 5-hydroxytryptamine [21] in the cortex, and indeed by NO (see Chapter 7). The NMDA receptor has become the workhorse mechanism linking the cellular level to the functional [most commonly long-term potentiation (LTP)] and to the dysfunctional (excitotoxicity). It seems hard to believe that only one such receptor complex with such versatility exists; a probable development in the future, then, will be identification of additional, comparable receptor complexes with analogous roles.

How do transmitter actions relate to function?

Long ago, it was fashionable to link proteins to specific functions, such as memory, and indeed to regard transmitters as having highly specific functions. For example, it is still sometimes suggested that dopamine makes a direct one-to-one contribution to a complex state, such as pleasure. Although such thinking is now being curtailed as findings accumulate showing that any one transmitter can participate in all number of brain functions, the advent of molecular biology has, in certain cases, renewed confidence in relating cellular events once more on a one-to-one basis with function (see Chapter 8) and with dysfunction (see Chapter 12). Moreover, the use of knock-out and transgenic mice has appeared to lead to the identification of, for example, the genetic basis of the NMDA receptor, a prerequisite for successful LTP and thus for the formation of spatial memory (see Chapter 13). Intriguing though this type of work is, however, it does not necessarily indicate a single causal relationship between the initial molecular phenotype and the final behaviour observed. Note that, for example, different gaseous transmitters are also needed in different ways for different types of LTP (see Chapter 7). Perhaps a more helpful way forward, as Huang and Stevens (Chapter 13) suggest, is to try and appreciate how a particular phenotype at the cellular level contributes to the nested hierarchy of brain organization, as its cellular significance becomes diluted in the progression from cells to synapses, extrajunctional contacts and dendrites, to circuits of ever more complexity, to brain regions, and eventually to net global brain organization. This issue of context is highlighted further by Tear (see Chapter 1): other bioactive molecules, trophic and tropic substances, can exert differential actions according to additional factors, such as disposition and variation in potential target sites within particular neuronal circuitry.

One way in which we will, in the future, increasingly be able to appreciate how neuronal mechanisms contribute to larger-scale functions within the brain is by imaging. In the past, the classical view was that one neuron, or a very small committed group of neurons, would respond to a visual stimulus in a predictable and highly localized way. Now, however, imaging of the visual cortex, using voltage-sensitive dyes [22], has revealed that many more neurons are involved over a time and space resolution that would previously have been impossible to explore. For example, when a light is shone at a monkey, no less than 10 million cells can be activated! In contrast with predictions from standard electrophysiological experiments, in which recording is at most from 100 cells, but most usually from only one or two, Grinwald et al. [22] have demonstrated that activity spreads following dendritic integration at speeds of some 250 μm/s.

In humans, imaging techniques cannot be invasive, so current methods, such as functional magnetic resonance imaging (MRI) and positron emission tomography (PET), exploit the variable expenditure of energy, namely the differential consumption of glucose and oxygen from one region to another. Unfortunately, these windows on the conscious human brain have a resolution currently only, at best, in the second time-scale. Nonetheless, PET can already yield valuable information on longer-lasting contributory factors, such as drug action, in dysfunctions such as schizophrenia (see Chapter 9). A safe prediction for the future is that the time resolution using, say, functional MRI will be brought into the subsecond time-scale. Then we will truly start to see how events at the level of the single cell contribute and build up to more global activation in the brain, and in turn how such global and highly transient neuronal configurations relate to the subtle functions of human cognitive function and dysfunction.

How do transmitters relate to dysfunction?

Just as a single cellular mechanism could not be responsible for any single brain function, so it is becoming increasingly apparent that dysfunctions cannot be attributed to the malfunction, exclusively, of any particular transmitter action at the cellular level. For example, schizophrenia has been associated, not so much with an excess of dopamine as with an aberrant interaction with other transmitters such as glutamate (see Chapter 9). Similarly, Parkinson's disease, for so long attributed solely to a loss of dopamine from the substantia nigra, is now linked to wider aberrant transmitter interactions in that region [23]. The other major neurodegenerative disorder, Alzheimer's disease, is, in a similar way, traditionally regarded as being a direct consequence of a loss of brain ACh. However, it is now widely accepted that other brain regions that are not cholinergic, such as the locus ceruleus and the raphe nuclei, also exhibit the pathological histology of neurofibrilliary tangles. Conversely areas of the brain that are cholinergic, such as the posterior pontine groups Ch 5 and 6, do not degenerate in Alzheimer's disease [24]. In the future, therefore, pharmaceutical strategies aimed at replacing ACh, similar to that of replacing dopamine using L-Dopa in Parkinson's disease, will not necessarily be the most desirable option.

One route towards a different approach lies in the development of knock-out and transgenic animals. For example, the development of transgenic animals expressing amyloid precursor protein will provide a valuable animal model for Alzheimer's disease [25]. On the other hand, it should be noted that manipulation of amyloid precursor protein in animals still does not establish the causal train of events that lead to the degeneration seen in patients. Similarly, the overexpression of acetylcholinesterase in transgenic mice, who accordingly display cognitive impairments reminiscent of Alzheimer's disease [26], does not in itself inspire an improved therapeutic strategy. Rather, such a model demonstrates the importance of acetylcholinesterase as a contributory factor, not least with regard to its non-cholinergic action (see above).

A second, alternative approach is to start with the cellular or molecular end-result of the disease, i.e. a pathological marker such as the amyloid deposits that characterize Alzheimer's disease (see Chapter 10) or mitochondrial dysfunction in the case of Parkinson's disease (see Chapter 4). The idea would then be to extrapolate back from the final common path, the proximal cause of cell death, to discover how such an aberration could have occurred. The problem, however, is that the end-result could, in turn, have been instigated by different factors, or several otherwise innocuous factors working in synergistic and pernicious conjunction [23].

A third, yet not mutually exclusive, way forward will be to attempt to understand what degenerative mechanism might be common to Alzheimer's and Parkinson's diseases. After all, the two pathologies can frequently co-exist and share a common, vulnerable target population of neurons in the brain – the very neurons that may make extensive use of promiscuous and diffuse dendritic signalling, as well as retaining developmental regenerative capacities. Could an ability for regeneration (see Chapter 5) retained selectively into adulthood be linked to the fact that it is these very neurons that are particularly prone to degeneration? Just such an association between mechanisms of development (see Chapter 1) and degeneration has already been proposed in relation to Alzheimer's disease [24].

Conclusions

We are about to enter an exciting phase in brain research, where there is a shift in emphasis away from the all-pervasive paradigm of classical synaptic transmission. In the light of recent findings, the familiar communication via the synapse should be regarded more as a cornerstone than as the complete foundation of brain function. Much signalling between cells is not local, not simply inhibitory or excitatory, and not a fast one-off event. By understanding more about non-classical neuronal mechanisms and previously undreamed of properties and processes, such as those described in the preceding essays, we can start to appreciate the full repertoire of neurochemical systems available to the brain, and begin to understand how aberrations in these mechanisms might account for certain dysfunctions. Molecular biology techniques will prove very powerful in helping us to establish the links, albeit indirect ones, between the cellular and molecular level of brain operations and that of cognitive functions and dysfunctions. But it is important not to make the mistake of assuming a simplistic one-to-one relationship between a type of behaviour and a gene or molecular phenotype. Rather, molecular biology has its place in the arm oury of awesome new techniques, such as fast-scan cyclic voltammetry and optical imaging, that should enable us, finally, to appreciate both the remoteness of a local neuronal event from final brain function as well as its indispensability to that function.

Summary

- *Neuronal signalling cannot always be reduced to classical synaptic transmission resulting in a binary inhibition or excitation.*
- *Many types of molecules that are bioactive do not fit readily into the profile of a classical neurotransmitter.*
- *Release of bioactive substances can occur beyond the familiar axonal synapse.*
- *The actions of bioactive substances, including classical transmitters, can be modulatory, global and long term.*
- *It is erroneous to extrapolate from events at the level of the neuron or gene to wholescale functions and dysfunctions.*
- *The actions of bioactive substances have to be appreciated within the context of neuronal populations, and in turn within the holistic anatomical context of the brain.*
- *Molecular biology will contribute to this more precise understanding of the relationship between individual genes, gene mutations, molecular and cellular phenotypes, and corresponding functions and dysfunctions.*

References

1. Hokfelt, T. (1991) Neuropeptides in perspective: the last ten years. *Neuron* **7**, 867–879
2. Greenfield, S.A. (1991) A noncholinergic action of acetylcholinesterase (AChE) in the brain: from neuronal secretion to the generation of movement. *Mol. Cell. Neurobiol.* **11**, 55–77
3. Agnati, L.F., Zoli, M., Stromberg, I. & Fuxe, K. (1995) Intercellular communication in the brain: wiring versus volume transmission. *Neuroscience* **69**, 711–726
4. Cheramy, A., Leviel, V. & Glowinski, J. (1981) Dendritic release of dopamine in the substantia nigra. *Nature (London)* **289**, 537–542
5. Cragg, S.J., Rice, M.E. & Greenfield, S.A. (1997) Heterogeneity of electrically evoked dopamine release and reuptake in substantia nigra, ventral tegmental area, and striatum. *J. Neurophysiol.* **77**, 863–873
6. Rice, M.E., Cragg, S.J. & Greenfield, S.A. (1997) Characteristics of electrically evoked somatodendritic dopamine release in substantia nigra and ventral tegmental area in vitro. *J. Neurophysiol.* **77**, 853–862
7. Cole, A.E. & Nicoll, R.A. (1983) Acetylcholine mediates a slow synaptic potential in hippocampal pyramidal cells. *Science* **221**, 1299–1301
8. Cragg, S.J. & Greenfield, S.A. (1997) Differential autoreceptor control of somatodendritic and axon terminal dopamine release in substantia nigra, ventral tegmental area and striatum. *J. Neurosci.* **17**, 5738–5746
9. Llinas, R.R. (1988) The intrinsic electrophysiological properties of mammalian neurons: insights into central nervous system function. *Science* **242**, 1654–1664
10. Anderson, P., Storm, J. & Wheal H.V. (1987) Thresholds of action potentials evoked by synapses on the dendrites of pyramidal cells in the rat hippocampus *in vitro*. *J. Physiol. (London)* **383**, 509–526
11. Nedergaard, S., Bolam, J.P. & Greenfield, S.A. (1988) Facilitation of a dendritic calcium conductance by 5-hydroxytryptamine in the substantia nigra. *Nature (London)* **333**, 174–177
12. Greenfield, S.A. (1985) The significance of dendritic release of transmitter and protein. *Neurochem. Int.* **7**, 887–901
13. Cragg, S.J., Holmes, C., Hawkey, C.R. & Greenfield, S.A. (1998) Synchronous, oscillatory dopamine release from developing midbrain neurons in organotypic culture. *Neuroscience*, **84**, 325–330
14. Woolf, N.J. (1996) Global and serial neurons form a hierarchically arranged interface proposed to underlie memory and cognition. *Neuroscience* **74**, 625–651
15. Levey, A.I., Hersch, S.M., Rye, D.B., Sunahara, R.K., Niznik, H.B., Kitt, C.A., Price, D.L., Maggio, R., Brann, M. R., Ciliax, B.J. et al. (1993) Localization of D1 and D2 dopamine receptors in brain with subtype-specific antibodies. *Proc. Natl. Acad. Sci. U.S.A.* **90**, 8861–8865
16. Hosokawa, T., Rusakov, D.A., Bliss, T.V. & Fine, A. (1995) Repeated confocal imaging of individual dendritic spines in the living hippocampal slice: evidence for changes in length and orientation associated with chemically induced LTP. *J. Neurosci.* **15**, 5560–5573
17. McCormick, D.A. & Prince, D.A. (1986) Acetylcholine induces burst firing in thalamic reticular neurones by activating a potassium conductance. *Nature (London)* **319**, 402–405
18. Crunelli, V. & Leresche, N. (1991) A role for GABAb receptors in excitation and inhibition of thalamocortical cells. *Trends Neurosci.* **14**, 16–21
19. Gray, R. & Johnston, D. (1987) Noradrenaline and β-adrenoreceptor agonists increase activity of voltage-dependent calcium channels in hippocampal neurons. *Nature (London)* **327**, 620–622
20. Nyrasem, J., Ryu, P.D. & Randic, M. (1986) Substance P augments a persistent slow inward calcium-sensitive current in voltage clamped spinal dorasal horn neurons of the rat. *Brain Res.* **365**, 369–376
21. Nedergaard, S., Flatman, J.A. & Engberg, I. (1991) The modulation of excitatory amino acid responses by serotonin in the cat neocortex *in vitro*. *Mol. Cell. Neurobiol.* **7**, 367–379

22. Grinwald, A., Lieke, E.E., Frostrig, R.D. & Hildesheim, R. (1994) Cortical point spread and long
     range lateral interactions revealed by real-time optical imaging of macaque monkey primary visual
     cortex. *J. Neurosci.* **14**, 2545–2568

23. Greenfield, S.A. (1992) Cell death in Parkinson's disease. *Essays Biochem.* **27**, 103–118

24. Greenfield, S.A. (1996) Non-classical actions of cholinesterases: role in cellular differentiation,
     tumorigenesis and Alzheimer's disease: a critique. *Neurochem. Int.* **28**, 485–490

25. Yamatsuji, T., Okamoto, T., Takeda, S., Murayama, Y., Tanaka, N. & Nishimoto, I. (1996)
     Expression of V642 APP mutant causes cellular apoptosis as Alzheimer trait-linked phenotype.
     EMBO J. **15**, 498–509

26. Beeri, R., Andres, C., Lev-Lehman, E., Timberg, R., Huberman, T., Shani, M. & Soreq, H. (1995)
     Transgenic expression of human acetylcholinesterase induces progressive cognitive deterioration
     in mice. *Curr. Biol.* **5**, 1063–1071

Subject index

A

Aβ42(43), 123–125
acetylcholine, 182, 185
acetylcholine receptor, 23
Alzheimer's disease, 46, 47–48, 117–131, 188
amine, 183
γ-aminobutyric acid (GABA), 182
amygdala, 176
amyloid cascade hypothesis, 119
amyloid deposition, 118
amyloid precursor protein, 118–123
analgesia, 68, 75
androgen receptor gene, 157
anticipation, 150
anti-psychotic drug, 109–110, 113–114
Aplysia, 61
apolipoprotein E, 126–127
apoptosis, 43, 49–50, 56
apoptosis-activating factor-1, 49
apoptosis-inducing factor, 49
L-arginine, 80
ataxin, 158, 160
autosomal dominant Alzheimer's disease, 118, 125–126
axonal guidance, 2, 9–11
axotomy, 55

B

blood–brain barrier, 135–136
bovine spongiform encephalopathy, 133
brain graft, 135–136

C

Ca^{2+}/calmodulin-dependent kinase, 170
CAG trinucleotide repeat disease, 156

calcium
  channel, 32
  sensor, 35
CAP-23, 61
carbon monoxide, 79
cell adhesion molecule, 8
chemoattraction/chemorepulsion, 2
chromatolysis reaction, 55
clozapine, 111
commissureless, 9
Cre, 171
Creutzfeldt–Jakob disease, 133
cyclic AMP, 172–173
cyclic AMP reporter cell line, 22
cyclic AMP-response-element-binding protein, 174
cyclic GMP, 82
cytochrome *c*, 50
cytoskeleton, 38

D

dentatorubral–pallidoluysion atrophy, 152, 158, 160
dopamine, 106, 111, 187
dopamine receptor, 106, 107, 111, 185
dorsal root ganglion, 57
Drosophila
  gene, 173
  memory, 172–174
  protein, 9
drug action on schizophrenia, 109–113
dunce, 173

E

endothelium-derived relaxing factor, 79
Eph family, 7

ephrin, 7
N-ethylmaleimide-sensitive factor (NSF), 30
expression screening, 18
extracellular matrix, 8

F

fast-scan cyclic voltammetry, 183, 184
fragile X syndrome, 150, 151
Friedreich's ataxia, 150, 151

G

GAP-43, 56, 59, 61
genetic linkage analysis, 108–109
glutamate, 108, 182
G-protein, 72, 97
G-protein-coupled receptor, 15, 16, 24–25,
    71, 94–95
growth-associated protein, 56
growth cone, 1
GTP-binding protein, 10, 36
guidance molecule, 9

H

haem oxygenase, 80–82
head trauma, 45
heroin, 65
hippocampus, 168
homology cloning, 17
huntingtin, 155, 159, 160
Huntington's disease, 46, 48, 149–163
5-hydroxytryptamine, 108
5-hydroxytryptamine receptor, 111
hyperalgesia, 88

I

immunoglobulin cell adhesion molecule, 8
inflammation, 88
inositol polyphosphate, 37
intermediate target, 9
inverse agonism, 21
ischaemia, 45

K

'knock-out' gene
    neurotransmitter receptor, 20
    nitric oxide synthase, 85, 89
    $Prnp^{\%}$, 138, 140

L

learning, 54
leucine-rich acidic nuclear protein, 158
ligand binding, 23–24, 110
ligand-gated ion channel, 15, 16
long-term memory, 174, 175
long-term potentiation, 83–88, 165, 167–170,
    186
loxP, 171

M

Machado–Joseph disease, 152
magnetic resonance imaging, 187
membrane fusion, 33
memory, 54, 165, 172–176
1-methyl-4-phenylpyridinium, 47
mitochondrial DNA mutation, 48–49
mono-ADP-ribosyltransferase, 87
morphine, 65
mRNA editing, 19
muscarinic acetylcholine receptor, 23
myotonic dystrophy, 150, 151
myristoylation, 81

N

nerve sprouting, 60–62
netrin, 3–5
neural connectivity, 1
neuroactive peptide, 181
neuroectodermal graft, 135
neuro-endocrine physiology, 68
neurograft for prion research, 137
neuronal
    differentiation, 53
    inhibition and excitation, 179

neuronal nitric oxide synthase, 85

neuropilin, 5

neuro-regeneration, 53–64

neurotoxin, 34

neurotransmitter, 179, 180

nitric oxide, 79

nitric oxide-independent long-term
    potentiation, 86

nitric oxide synthase, 80–82, 85, 89

N-methyl-D-aspartate (NMDA) receptor,
    168, 186

NSF (see N-ethylmaleimide-sensitive factor)

nuclear inclusion, 157–160

O

oculopharyngeal muscular atrophy, 153

odorant complementarity-determining region,
    96

odorant-dependent signal transduction, 97

Odr-10, 99

olfactory
    bulb, 101
    G-protein, 97
    pathway, 93
    receptor, 93–104

opiate action, 65–77

opioid
    peptide, 67, 68
    receptor, 67, 70, 74

orphan receptor, 18

P

palmitoylation, 31

Parkinson's disease, 46–47, 59, 187, 188

peripheral inflammation, 88

phosphoinositide, 35

phosphorylation, 38

polyglutamine expansion disease, 149–163

polymorphic variation, 19

positron emission tomography, 112, 187

presenilin, 118–123

prion
    disease, 133–147
    in central nervous system, 138–141

neurograft technique, 137
    protein, 134, 138, 140, 141

Prnp$^{\%}$ mouse, 138, 140

programmed cell death, 43, 49–50, 56

progressive myoclonus epilepsy, 151

protein kinase, 169

proteolytic processing, 160

proton motive force, 45

PrP^{C}, 134, 138, 140, 141

PrP^{Sc}, 134, 138, 141

R

Rab family, 36

receptor
    androgen, 157
    architecture, 23–25
    dopamine, 106, 107, 111, 185
    G-protein-coupled, 15, 16, 25, 71,
        94–95
    5-hydroxytryptamine, 111
    ligand-gated ion channel, 15, 16
    muscarinic acetylcholine, 23
    N-methyl-D-aspartate, 168, 186
    olfactory, 93–104
    opioid, 67, 70, 74
    receptor–G-protein coupling, 24–25
    SNAP receptor (SNARE), 30
    UNC-5, 5

recombinant expression, 21

reporter cell line, 22

respiratory chain inhibitor, 44

retrograde messenger, 83

Rho family, 10

rotenone, 47

roundabout, 9

S

schizophrenia, 105–116, 187

scrapie, 133, 136, 143

semaphorin, 5

serotonin, 108

signal transduction, 9–10, 97

single-photon emission tomography scanning,
    112

small GTP-binding protein, 10, 36
small synaptic vesicle, 29, 34
SNAP (soluble *N*-ethylmaleimide-sensitive-
     factor accessory protein), 30
SNAP-25, 31
SNARE (SNAP receptor), 30
spinal and bulbar muscular atrophy, 152, 157
spinal cord, 88–89
spinocerebellar ataxia, 150, 152, 155
splice variant, 18
spongiform encephalopathy, 133, 141–145
stress-induced analgesia, 68
synapsin, 38
synaptic plasticity, 53, 167
synaptic transmission, 29
synaptic vesicle endocytosis, 39
synaptobrevin, 31
synaptogenesis, 60–62
synaptotagmin, 35
synpolydactily, 153

syntaxin, 31

T
tau protein, 127

tga20, **143**
therapeutic agent, 75
toxic property gain, 157
transglutaminase, 160
trinucleotide repeat expansion, 149

U
ubiquitin, 157, 158
UNC-5, 5–6

V
vesicle-associated membrane protein (VAMP),
     31
visual system, 54
VRSQ motif, 119